M000220341

EXPANSIVE SOILS

EXPANSIVE SOILS

Problems and Practice in Foundation and Pavement Engineering

John D. Nelson

Debora J. Miller

Department of Civil Engineering
Colorado State University

JOHN WILEY & SONS, INC.

New York / Chichester / Brisbane / Toronto / Singapore

In recognition of the importance of preserving what has been
written, it is a policy of John Wiley & Sons, Inc., to have books
of enduring value published in the United States printed on
acid-free paper, and we exert our best efforts to that end.

Copyright © 1992 by John Wiley & Sons, Inc.

All rights reserved. Published simultaneously in Canada.

Reproduction or translation of any part of this work
beyond that permitted by Section 107 or 108 of the
1976 United States Copyright Act without the permission
of the copyright owner is unlawful. Requests for
permission or further information should be addressed to
the Permissions Department, John Wiley & Sons, Inc.

This publication is designed to provide accurate and
authoritative information in regard to the subject
matter covered. It is sold with the understanding that
the publisher is not engaged in rendering legal, accounting,
or other professional services. If legal advice or other
expert assistance is required, the services of a competent
professional person should be sought. *From a Declaration
of Principles jointly adopted by a Committee of the
American Bar Association and a Committee of Publishers.*

Library of Congress Cataloging in Publication Data:

Nelson, John D.
   Expansive soils : problems and practice in foundation and pavement
engineering / by John D. Nelson and Debora J. Miller.
       p.   cm.
   Includes bibliographical references and index.
   ISBN 0-471-51186-2
   1. Soil-structure interaction.   2. Swelling soils.
3. Foundations.   4. Pavements—Subgrades.   I. Miller, Debora J.
II. Title.
TA711.5.N5   1992
624.1'5136—dc20   91-42720
                                                          CIP

Printed in the United States of America

10   9   8   7   6   5   4   3   2   1

Printed and bound by Courier Companies, Inc.

To Darlene and Lou,
good friends and great spouses.

PREFACE

The phenomenon of swelling of soils has been known for about as long as the field of geotechnical engineering has been practiced. However, the practice of foundation engineering has its origins primarily in locations having deep deposits of soft clays such as the large coastal North American, European, and Mexican cities. As construction grew in arid regions such as the western United States and Canada, the Middle East, Australia, and Africa, the problems associated with swelling and shrinking soils began to receive more attention. However, most universities did not offer formal courses in foundation and pavement engineering on expansive soils. Thus, practicing engineers usually learned about expansive soils the "hard way," after buildings and pavements began to experience distress.

Over the past three decades, a significant amount of research began to focus on expansive soils. Several conferences and workshops were sponsored. There have also been a number of training courses regarding the subject. The need for a book of this type became evident, and several years ago the National Science Foundation sponsored an investigation to review the literature, investigate the state of practice on expansive soils, and prepare a document that can serve both as a reference guide for the practicing engineer and as a textbook for a course on foundations and pavements on expansive soils. This book represents that document.

The organization of the book is described in detail in the introductory chapter. It begins with discussions of the factors that cause swelling, and the factors of which engineers need to be aware during site investigation in order to identify expansive soils. Next the mechanics of unsaturated soils is introduced. Soil suction is presented and discussed in detail because of the importance of this concept in regard to expansive soils. After that, methods of testing and applications to design of foundations and pavements are covered. Soil treatment to reduce swelling is discussed to the extent possible within the scope of this book. Finally, a discussion of remedial measures and principles behind design of remedial measures is presented. Although the discussion of remedial measures is not exhaustive, we believe that it represents the most current discussion of that topic at this time.

The investigation leading to this book was sponsored by the National Science Foundation, and was conducted at Colorado State University. The scope of the investigation included a complete review of the literature and preparation of an annotated bibliography. It also included a major effort devoted to visiting and interviewing practicing engineers engaged in work involving expansive soils. These included engineers from private practice, government, research organizations, and academia. Visits were made to engineers in Saskatchewan, Canada, and most of the western states and several midwestern states in the United States. The companies and organizations visited are mentioned in the book. The material in the book also draws on personal experience in South Africa and Israel. After collection of the data and the literature review, a draft document was prepared and distributed to 29 experts on expansive soils from private practice, government, research, and academia. A two-day meeting was convened at which time the workshop experts reviewed the draft document in detail, made recommendations for additions and revisions, and helped to develop a detailed outline for the book. The members of this panel of experts are listed in the acknowledgments.

Although the panel made very detailed comments and suggestions, the opinions expressed in the book should be attributed only to the authors. The lapse of a significant amount of time between the meeting and the publication of the book has necessitated significant deviation from the conclusions expressed at the panel meeting. For those variations the authors accept full responsibility.

The NSF project also included a public awareness and training component. A booklet entitled "Building on Expansive Soils," by Janis L. Fenner, Debora J. Hamberg (now Miller), and John D. Nelson, was published by Colorado State University. A series of half-day seminars were presented around the western United States at 14 different cities for engineers, contractors, real estate personnel, lawyers, and other professionals involved in expansive soils.

JOHN D. NELSON
DEBORA J. MILLER

Fort Collins, Colorado
October 1991

ACKNOWLEDGMENTS

The investigation that led to this book was sponsored by the National Science Foundation (NSF) of the United States. The book was not completed during the NSF grant time period, and Colorado State University continued to support preparation of the manuscript both financially and with encouragement. John D. Nelson served as the Principal Investigator on the project. Debora J. Miller worked on the NSF project as a graduate student, and she continued as a coauthor after she received her degree. During the NSF grant period, several graduate and undergraduate students assisted with the project in many different aspects. These students included Joseph C. Goode, Janis L. Fenner, Thomas V. Edgar, Joseph P. Martin, Andrew Porter, Louis L. Miller, Hansreudi Schneider, Michelle Howard, and Connie Madrid. In addition, Dr. Amos Komornik shared his experiences relating to practice in Israel.

The authors also want to acknowledge the individuals who attended the workshop held at Colorado State University in December 1982, and contributed recommendations for content and detailed outlines for each chapter of this book. These experts included James C. Armstrong, Ernest Buckley, Fu Hua Chen, Wayne Clifton, Joseph C. Goode, Gordon Eischens, Joan D. Finch, Delwyn G. Fredlund, Stephen S. Hart, Wesley G. Holtz, Lawrence D. Johnson, Earl Jones, Paul Knodel, Leonard Krazynski, Robert Lytton, John McCabe, R. Gordon McKeen, Return Moore, Michael O'Neill, Thomas M. Petry, Michael J. Rose, John Schmertmann, Joseph Sheffield, Donald R. Snethen, Malcolm Steinberg, Bill Stroman, Merrill Walstad, Don Weichlen, and Warren K. Wray.

The authors are also very grateful to Dr. Michael P. Gaus and Dr. J. Eleonora Sabadell. Dr. Gaus was on the staff of the National Science Foundation at the time of initiation of the grant. Dr. Sabadell continued as the NSF project monitor after Dr. Gaus retired from the NSF. Without their insight, patience, and encouragement, completion of the manuscript would not have been possible.

Finally we owe a debt of gratitude to Constance Atkins, Laurie J. Howard, Kenneth H. J. Streeb, and Sandy Wittler for typing the manuscript and drafting of the figures. All of these individuals went out of their way to provide excellent help in a very timely way.

CONTENTS

LIST OF SYMBOLS

SYMBOL		**REFERENCE**
COLE	= coefficient of linear extensibility	Ch. 3, 4
CPT	= cone penetration test	Ch. 2
CS	= consolidation-swell test	Ch. 5
CV	= constant volume test	Ch. 5
c	= soil cohesion	Ch. 5
D_m	= water content index [slope of $w - \log(u_a - u_w)$ relationship]	Ch. 4
D_t	= water content index [slope of $w - \log(\sigma - u_a)$ or $w - \log(\sigma - u_w)$ relationship]	Ch. 4
d	= beam depth	Tbl. 5.6
d	= pier diameter	Ch. 5
d_b	= diameter of pier bell	Ch. 5
E	= edge distance, distance from edge of slab to point of contact with mound	Fig. 5.14
E	= modulus of elasticity of soil	Ch. 5
E_D	= dilatometer modulus	Ch. 2
E_s	= modulus of elasticity of soil	Ch. 5
EI	= expansion index	Ch. 3, 5
e	= void ratio	Ch. 2
e	= edge moisture variation distance	Ch. 5
e_f	= final void ratio	Ch. 4
e_o	= initial void ratio	Ch. 4
e_m	= edge moisture variation distance	Fig. 5.17
F	= van der Merwe reduction factor	Ch. 4
f	= slab-subgrade friction coefficient	Tbl. 5.6
f_s	= skin friction below active zone	Ch. 5
f_u	= frictional stress between soil and pier	Ch. 5
f_u	= pier-soil strength	Ch. 5
f_{um}	= maximum pier-soil strength at tip	Ch. 5
G_s	= specific gravity of solids	Ch. 4
g	= gravitational acceleration	Ch. 4
H	= height above water table	Fig. 4.22
h	= total suction head	Ch. 4
h	= total soil depth	Ch. 5
h_c	= matric suction head	Ch. 4
h_d	= displacement pressure head	Ch. 4
h_o	= osmotic suction head	Ch. 4
h_m	= matric suction	Tbl. 4.1
h_s	= solute suction	Tbl. 4.1
I	= dimensionless parameter for pier movement	Ch.5
I_D	= material index in Dilatometer test	Ch. 2

SYMBOL

I_{pt}''	= total suction index	Tbl. 4.2
I_{pm}''	= matric suction index	Tbl. 4.2
I_{ps}''	= solute suction index	Tbl. 4.2
K_D	= horizontal stress index in Dilatometer test	Ch. 2
K_o	= coefficient of earth pressure at rest	Ch. 2
k	= subgrade modulus	Ch. 5
k	= factor between 0 and 1	Eq. 5.5 Ch. 5
L	= linear dimension	Ch. 3, 4
L	= pier length	Eq. 5.4 Ch. 5
L	= slab length	Tbl. 5.6
L_s	= linear shrinkage	Ch. 3
LE	= linear extensibility	Ch. 3
LL	= liquid limit	Ch. 2, 3
LMO	= lime modification optimum	Ch. 6
m	= mound shape exponent	Ch. 5
N_u	= parameter in determination of bell uplift resistance	Eq. 5.6
n	= porosity	Fig. 4.22
P	= load on pier	Ch. 5
P_b	= load at tip of pier	Ch. 5
P_{dl}	= pier dead load	Eq. 5.5 Ch. 5
P_{FS}	= maximum load mobilized by skin friction	Ch. 5
P_{max}	= maximum load in pier	Ch. 5
PE	= potential expansiveness (van der Merwe method)	Ch. 4
PI	= plasticity index	Ch. 2, 3
PL	= plastic limit	Ch. 2, 3
PVC	= potential volume change	Ch. 3
PVR	= potential vertical rise	Ch. 5
p'	= pore water pressure deficiency	Tbl. 4.1
p_o'	= initial effective overburden stress	Ch. 2
pF	= $\log_{10}$ (suction head in cm H_2O)	Ch. 2
Q_u	= uplift resistance of pier bell	Eq. 5.6 Ch. 5
q	= applied bearing pressure	Ch. 5
q_s	= allowable soil bearing pressure	Ch. 5
q_c	= static cone resistance	Ch. 2
q_c	= center load on slab	Tbl. 5.6
q_{dl}	= unit dead load pressure	Eq. 5.4 Ch. 5
q_c	= edge load on slab	Tbl. 5.6
R	= shrinkage ratio	Ch. 3
R	= net interparticle repulsive force	Tbl. 4.1

SYMBOL   REFERENCE

R	= universal gas constant	Ch. 4
r	= radius	Ch. 4
S	= degree of saturation	Fig. 4.22
S	= beam spacing	Tbl. 5.6
S_p	= percent swell (Schneider and Poor method)	Ch. 4
S_r	= residual saturation	Ch. 4
S_u	= undrained shear strength	Ch. 2
SI	= swell index	Ch. 3
SL	= shrinkage limit	Ch. 3
SPT	= standard penetration test	Ch. 5
T	= absolute temperature	Ch. 4
T_s	= surface tension	Ch. 4
TMI	= Thornthwaite Moisture Index	Ch. 2
U	= uplift force on pier	Ch. 5
u_a	= pore air pressure	Ch. 2
u_w	= pore water pressure	Ch. 2
V	= volume	Ch. 3
W	= withholding force	Ch. 5
w	= water content	Ch. 2
w	= average pressure under slab	Tbl. 5.6
w_l	= liquid limit	Tbl. 5.9
w_n	= natural water content	Tbl. 5.9
w_p	= plastic limit	Tbl. 5.9
Y	= coordinate defining mound profile under slab	Ch. 5
Y_m	= maximum mound height under slab	Fig. 5.14
Y_{max}	= maximum mound height with no load	Fig. 5.14
z	= vertical distance from ground water table	Ch. 2
z	= thickness of expansive soil layer	Ch. 5
z	= depth below ground surface	Ch. 5
z_a	= depth of active zone	Ch. 5
z_i	= thickness of incremental layer	Ch. 4
A, α Alpha		
α	= compressibility factor	Tbl. 4.2
α_1	= coefficient of uplift between pier and soil	Ch. 5
α_2	= correction factor for skin friction	Ch. 5
B, β Beta		
β	= statistical factor related to contact area	Tbl. 4.1

SYMBOL		**REFERENCE**
β	= parameter related to pier and bell diameters	Eq. 5.7 Ch. 5
β'	= bonding factor	Tbl. 4.1
Γ, γ Gamma		
γ	= soil unit weight	Eq. 5.6 Ch. 5
γ_d	= dry unit weight (dry density)	Ch. 3,4
γ_{dD}	= dry density of oven dry sample	Eq. 3.3 Ch. 4
γ_{dM}	= dry density of sample at 33kPa suction	Eq. 3.3 Ch. 4
γ_h	= total suction index	Tbl. 4.2
γ_t	= total unit weight	Ch. 2
γ_w	= unit weight of water	Ch. 2
γ_{sat}	= saturated unit weight	Ch. 4
Δ, δ Delta		
Δ	= differential movement	Tbl. 5.7
E, ϵ Epsilon		
ϵ_v	= vertical strain	Tbl. 4.2
Θ, θ Theta		
θ	= volumetric water content	Fig. 4.22
M, μ Mu		
μ	= matric suction (pressure units)	Ch. 2, 4
μ	= microns	Ch. 3
N, ν Nu		
ν	= Poisson's ratio	Ch. 5
ν_s	= Poisson's ratio of soil	Ch. 5
P, ρ Rho		
ρ	= mass density	Ch. 4
ρ	= settlement (or heave)	Ch. 4
ρ_m	= maximum heave under slab	Ch. 5
ρ_{max}	= maximum total predicted heave	Ch. 5
ρ_p	= movement of pile at surface	Ch. 5
ρ_o	= free field heave at surface	Ch. 5
Σ, σ Sigma		
σ	= total normal stress	Ch. 2
σ'	= effective normal stress	Ch. 2
σ'_c	= matric suction equivalent	Fig. 4.17 Ch. 4
σ'_f	= effective stress at final conditions	Ch. 4, 5
σ_h	= total horizontal stress	Ch. 2
σ'_h	= swelling pressure in terms of effective stress	Ch. 4, 5
σ'_{sc}	= corrected swelling pressure	Ch. 4
σ_v	= total vertical stress	Ch. 2

SYMBOL		REFERENCE
$\sigma_1, \sigma_2, \sigma_3$	= principle stresses	Ch. 2
T, τ Tau		
τ	= shear stress	Ch. 2
Φ, ϕ Phi		
ϕ_r'	= residual angle of internal friction	Ch. 5
X, χ Chi		
X	= Chi parameter	Tbl. 4.1
Ψ, ψ Psi		
ψ	= parameter ranging from zero to one	Tbl. 4.1
Ω, ω Omega		
Ω	= osmotic pressure	Ch. 4

1

INTRODUCTION

Expansive soil is a term generally applied to any soil or rock material that has a potential for shrinking or swelling under changing moisture conditions. When a soil is referred to in this book as *expansive* or when reference is made to *swell potential* it should be recognized that there also exists a potential for shrinking or settlement to occur due to changes in moisture content. Thus, the terms *expansive soil* and *swell potential* will generally be used in a universal sense to refer to soils that both shrink and swell. This does not include, however, the class of soils generally referred to as *collapsible* soils.

The primary problem that arises with regard to expansive soils is that deformations are significantly greater than elastic deformations and they cannot be predicted by classical elastic or plastic theory. Movement is usually in an uneven pattern and of such a magnitude as to cause extensive damage to the structures and pavements resting on them.

Expansive soils cause more damage to structures, particularly light buildings and pavements, than any other natural hazard, including earthquakes and floods (Jones and Holtz, 1973). In a special report published by the Federal Emergency Management Agency, the following figures were cited (FEMA, 1982).

Selected annual U.S. losses from expansive soils were $798.1 million in 1970 and are expected to rise to $997.1 million by the year 2000 (Wiggins et al., 1978; Petak et al., 1978). These values are for *residence losses only*. Loss figures were developed on a consistent base using constant 1970 replacement dollars.

The cost of damage to other structures such as commercial/industrial buildings and transportation facilities raises these total estimated values by a factor of 2 to 3.

The Department of Housing and Urban Development (Jones, 1981) estimated average annual losses due to shrink–swell phenomena as $9 billion in 1981. A rank-ordered

1

list of hazards showed shrink–swell phenomena to be second, below insect damage, as the most likely natural hazard to cause economic loss. Jones' estimates were based partially on subjective judgements and, in part, on documentation. Documented evidence is limited, especially since this problem is not generally considered for official disaster declaration by federal or state agencies and losses are not generally covered by insurance. (An exception to this was the 1980 disaster declaration in Tulsa, Oklahoma, where extended drought caused shrinkage damage to residences in that area to such an extent that relief funding was made available through the Small Business Administration.)

Although the estimation of damage is inexact and somewhat subjective, even the most conservative estimates show that expansive soils are a major contributor to the burden that natural hazards place on the economy. Despite the fact that billions of dollars in yearly damage losses have been attributed to these problem soils, the state of the practice in design and construction is severely limited by continued lack of understanding of expansive soil behavior and soil–structure interaction. Also, a greater appreciation of risks associated with building on expansive soils must be developed on the part of owners, lending institutions, regulatory officials, builders, architects, and engineers. Loss mitigation, through responsible engineering and construction, is essential in helping to alleviate the risk from this natural hazard. The first steps in mitigation are to recognize the problem and understand the preventive options that are available. The next step, which is essential, is to provide careful, responsible engineering and construction. Research is also important, as it provides knowledge and data for input into understanding of phenomena and design.

1.1 PURPOSE

The purpose of this book is to provide a source of information for concerned individuals involved in all phases of investigation, design, construction, marketing, ownership, and repair of structures built on expansive soils. Techniques available for dealing with the problems associated with design and construction are presented, and reliable methods are pointed out. However, in each situation there will continue to be a degree of uncertainty. The current technology is not at a stage where a guarantee of "absolutely no damage" can be assured by design and construction practices. Through good understanding of the problems and phenomena, engineers should be capable of addressing the differences in risk associated with alternative solutions and communicating level of risk to their clients.

1.2 ORGANIZATION

The book is organized to follow, to the extent possible, the chronology of events and potential stages of investigation and design in the life of a structure founded on expansive soils.

Chapter 2 reviews the factors that affect swelling and shrinking of soils, and describes the site characterization process for assessing those factors. The factors affecting soil volume change are discussed in relation to soil properties, environmental effects, and state of stress. Methods of sampling expansive soils are discussed, and field techniques for evaluating expansive soil sites are summarized.

Laboratory identification tests and classification schemes are presented in Chapter 3. Many methods are available for reliably and inexpensively identifying soils that have expansion potential. Identification test procedures and classification schemes are summarized.

Chapter 4 presents methods for quantitatively predicting the amount of potential soil heave that can occur. Availability of water is a primary factor governing the potential magnitude and rate of soil movement. Therefore, environment must be considered in prediction of heave. Chapter 4 presents theoretical considerations, testing methods, and environmental factors that are important in heave prediction.

Design alternatives are presented in Chapter 5. Alternatives must be assessed according to many factors, such as site conditions, groundwater and soil characteristics, and costs. Advantages and disadvantages must be considered and an acceptable level of risk must be determined. In Chapter 5, alternatives related to both structural features and treatment are listed. Constraints associated with the alternatives are presented so that risks can be assessed for the conditions that exist. Foundation alternatives for buildings and treatment alternatives for highway subgrades are treated separately.

Soil treatment methods are discussed in Chapter 6. The appropriate type of treatment depends on the soil and environmental conditions, and the degree of risk the owner is willing to assume. Much research and some application of these techniques have been performed on federal government projects, in university studies, and in the construction industry. The recommendations made in this report regarding applicability and effectiveness of various methods of treatment under particular conditions are based on reports from these projects and interviews with consultants, builders, and investigators in academia and government.

Chapter 7 deals with remedial measures for structures that have been damaged by expansive soil movements. It is an important fact that remedial measures, such as releveling buildings or repaving highways and roads, typically cost far more than any initial "savings" that may have been realized from inadequate design and construction. Remedial work begins with a diagnosis of the problem. Diagnostic procedures are discussed. Example formats for diagnostic data gathering and reports are suggested and some remedial measures that have been used for different foundation types and for pavement systems are described and assessed for effectiveness and cost. The selection of a repair alternative must be based on an assessment of acceptable future risk. There frequently must be a compromise between the parties involved, particularly if those paying the repair costs are not those benefiting from the repairs. In many situations, there is disagreement about what constitutes failure. Some recommendations and comments about these aspects of acceptable risk are included in this section.

1.3 GENERAL CONSIDERATIONS

The major concern with expansive foundation soils in North America exists generally in the western and midwestern United States and Canada, and the Gulf States. In the course of preparing this book, consulting firms in a number of different locations in North America were visited. These include the Front Range area in Colorado; Tulsa, Stillwater, and Oklahoma City, Oklahoma; Dallas, Fort Worth, and Houston, Texas; Los Angeles and San Diego, California; Orlando and Gainesville, Florida; Minneapolis, Minnesota; Fargo and Minot, North Dakota; and Regina and Saskatoon, Saskatchewan, Canada.

Although expansive soils are encountered in nearly every state and province of the United States and Canada, the potential problems related to expansive soils in the northeastern areas are not as severe or widespread as problems encountered in the western and southern regions. In general, expansive soil problems may be categorized into different geographic areas. In the north-central and Rocky Mountain areas of North America, including North and South Dakota, Montana, Wyoming, Colorado, Alberta, and Saskatchewan, the expansive soil problems are primarily related to highly overconsolidated clays and weathered shales. In these areas, the foundation and pavement problems relate primarily to swelling as a result of increased moisture content after construction. In general, the time required for swelling to take place is relatively long, being on the order of 5 to 8 years. Rebound of these soils by unloading due to excavation may be another factor contributing to swell.

In eastern Oklahoma, eastern Texas, Florida and the Gulf States, and the Great Lakes and Lake Agassiz areas, the problem includes both shrinking and swelling. In those areas, moisture fluctuations over the different seasons cause shrinking during drier periods and swelling when they are wetted again. Volume change can take place over relatively short periods of time ranging from a few hours to many years.

In central and western Texas and western Oklahoma the problems include both swelling of overconsolidated, desiccated soils as well as shrinking. In Minnesota some expansive soil problems exist in the Cretaceous shales found primarily in the southeast and south central portions of the state.

In the southern California area the problem is primarily one of desiccated alluvial and colluvial soils. Many of the soils have a volcanic origin and have been desiccated in the natural state. Development construction is generally accompanied by irrigation around houses and buildings. The additional irrigation water causes swelling problems to occur. In this area, swelling can occur even after relatively short periods of time.

Foundation types range from shallow to deep foundations and the choice of a foundation depends to a large extent on the life styles and irrigation practices in the local area. Also, the amount of grading that is done during development of a site varies from place to place, depending on topography. In general, the warmer areas such as Texas, Florida, and California do not utilize basements under houses. In these areas, shallow foundations and slabs on grade are commonly used foundation systems. In the northern areas, basements have become common because of the need for furnaces and heating equipment. Deep foundations utilizing pier and grade beam construction are utilized more frequently in those areas.

Pavement systems, and canal and pipeline projects often utilize soil treatment methods. Large-scale applications of moisture barriers, lime and cement stabilization, and other techniques have been used by the state highway departments of Texas, New Mexico, Colorado, Mississippi, California, and Arizona. Large federal projects also have applied soil treatment methods for canals and for commercial and military airfields.

Site investigation and design practices vary from location to location, but within a particular area, local practices appear to be fairly similar to one another. Also, the degree of awareness on the part of the public and the engineering community varies from place to place. The engineering community is becoming more aware of the existence of expansive soils and their locations. As a result, more problems are diagnosed correctly, whereas in earlier years many expansive soils problems were incorrectly attributed to settlement.

Of major importance is the need to make owners aware of the potential problems that may exist and the fact that money spent on proper investigation and design is a worthwhile investment. In that regard the following quote is very appropriate (Krazynski, 1979):

> To come even remotely close to a satisfactory situation, trained and experienced professional geotechnical engineers must be retained to evaluate soil conditions. The simple truth is that it costs more to build on expansive soils and part of the cost is for the professional skill and judgement needed. Experience also clearly indicates that the cost of repairs is very much higher than the cost of a proper initial design, and the results are much less satisfactory.

2

SITE CHARACTERIZATION

The two major factors that must be identified in the characterization of a site for a building or a highway where potential shrink–swell problems may exist are

- the expansive or shrink–swell properties of the soil
- environmental conditions that contribute to moisture changes in the soil.

In any geotechnical site investigation, the subsoil profile and the physical properties of the subsurface materials must be investigated. If expansive soil exists, the environmental conditions that would contribute to moisture changes must also be evaluated and interpreted for their probable effects on swell potential. Environmental aspects are as important to expansive soil behavior as are the soil characteristics.

Normal soil investigation practice for nonexpansive sites is often not adequate in providing sufficient information to quantify the potential for expansion. If expansive soils are present, more extensive site investigation and specialized testing programs are justified, even for small structures such as houses and one-story buildings. For a large project, site investigation should be conducted in stages so as to optimize the use of funding and to enhance the amount of pertinent data that can be obtained.

2.1 ORGANIZATION OF INVESTIGATION

The three major phases of a staged investigation include reconnaissance, preliminary investigation, and final investigation. A flow chart of the site characterization process is shown on Figure 2.1.

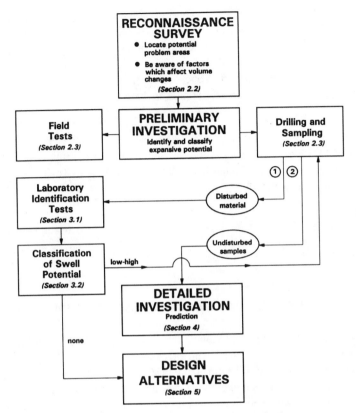

FIGURE 2.1. Flow chart for site characterization.

1. *Reconnaissance* is the first phase. It includes the use of maps, field or aerial observations, and local experience to locate potential problem areas. This information is then used to define the scope of the preliminary investigation.

2. *Preliminary investigation* is intended to confirm whether the soils at the site have a potential for swelling and shrinking. This investigation may include some preliminary subsurface sampling and initial laboratory testing and analysis. The subsurface profile should be defined as closely as possible, and soils should be identified and classified. Shrink–swell potential can be identified on the basis of various classification schemes. On the basis of the preliminary investigation a final, more detailed investigation can be defined.

3. *Detailed investigation* includes detailed definition of the soil profile, determination of soil properties, and quantification of the shrink–swell potential at the site. Quantitative site evaluation requires relatively undisturbed soil samples for prediction testing. Truly undisturbed samples of expansive soil are difficult to obtain and are generally expensive. However, with well trained and experienced drilling crews and good quality field control, standard sampling procedures can provide samples of reasonable quality.

The detailed investigation should be carefully planned. It is important to maintain flexibility so that the program can be modified to take advantage of data as they are gathered. For example, as the investigation progresses, the program of drilling and sampling should be reviewed and modified if necessary so as to fill in gaps and fully define suspected problem zones.

Important aspects of site characterization include

- The soil and environmental factors that affect shrink–swell behavior.
- The site exploration process, including techniques for drilling and sampling.
- The various methods available in the field for identifying expansive soils during preliminary investigation.

Laboratory identification tests and classification schemes are discussed in Chapter 3. Specialized testing methods for prediction of volume changes are covered in Chapter 4.

2.2 FACTORS INFLUENCING SWELLING AND SHRINKING OF SOILS

The mechanism of swelling in expansive clays is complex and is influenced by a number of factors. Expansion is a result of changes in the soil water system that disturb the internal stress equilibrium. Clay particles generally are platelets having negative electrical charges on their surfaces and positively charged edges. The negative charges are balanced by cations in the soil water that become attached to the surfaces of the platelets by electrical forces. The electrical interparticle force field is a function of both the negative surface charges and the electrochemistry of the soil water. van der Waals surface forces and adsorptive forces between the clay crystals and water molecules also influence the interparticle force field. The internal electrochemical force system must be in equilibrium with the externally applied stresses and capillary tension in the soil water. The capillary tension is often called matric suction.

If the soil water chemistry is changed either by changing the amount of water or the chemical composition, the interparticle force field will change. If the resulting change in internal forces is not balanced by a corresponding change in the externally applied state of stress the particle spacing will change so as to adjust the interparticle forces until equilibrium is reached. This change in particle spacing manifests itself as shrinkage or swelling.

Many of the factors influencing the mechanism of swelling also affect, or are affected by, physical soil properties such as plasticity or density. The factors influencing the shrink–swell potential of a soil can be considered in three different groups, the *soil characteristics* that influence the basic nature of the internal force field, the *environmental factors* that influence the changes that may occur in the internal force system, and the *state of stress*. The soil and environmental factors that affect swell and shrink behavior are summarized in Tables 2.1 and 2.2.

TABLE 2.1. Soil properties that influence shrink–swell potential

Factor	Description	References
Clay mineralogy	Clay minerals which typically cause soil volume changes are *montmorillonites*, *vermiculites*, and some *mixed layer minerals*. Illites and Kaolinites are infrequently expansive, but can cause volume changes when particle sizes are extremely fine (less than a few tenths of a micron)	Grim (1968); Mitchell (1973, 1976); Snethen et al. (1977)
Soil water chemistry	Swelling is repressed by increased cation concentration and increased cation valence. For example, Mg^{2+} cations in the soil water would result in less swelling than Na^+ cations	Mitchell (1976)
Soil suction	Soil suction is an independent effective stress variable, represented by the negative pore pressure in unsaturated soils. Soil suction is relaed to saturation, gravity, pore size and shape, surface tension, and electrical and chemical characteristics of the soil particles and water (see Chapter 4)	Snethen (1980); Fredlund and Morgenstern (1977); Johnson (1973); Olsen and Langfelder (1965); Aitchison et al. (1965)
Plasticity	In general, soils that exhibit plastic behavior over wide ranges of moisture content and that have high liquid limits have greater potential for swelling and shrinking. Plasticity is an *indicator* of swell potential	See Section 3.1
Soil structure and fabric	Flocculated clays tend to be more expansive than dispersed clays. Cemented particles reduce swell. Fabric and structure are altered by compaction at higher water content or remolding. Kneading compaction has been shown to create dispersed structures with lower swell potential than soils statically compacted at lower water contents	Johnson and Snethen (1978); Seed et al. (1962a).
Dry density	Higher densities usually indicate closer particle spacings, which may mean greater repulsive forces between particles and larger swelling potential	Chen (1973); Komornik and David (1969); Uppal (1965)

9

TABLE 2.2. Environmental conditions that influence shrink–swell potential

Factor	Description	References
1. Initial moisture condition	A desiccated expansive soil will have a higher affinity for water, or higher suction, than the same soil at higher water content, lower suction. Conversely, a wet soil profile will lose water more readily on exposure to drying influences, and shrink more than a relatively dry initial profile. The initial soil suction must be considered in conjunction with the expected range of final suction conditions	Johnson (1969)
2. Moisture variations	Changes in moisture in the active zone near the upper part of the profile primarily define heave. It is in those layers that the widest variation in moisture and volume change will occur.	
2.1 Climate	Amount and variation of precipitation and evapotranspiration greatly influence the moisture availability and depth of seasonal moisture fluctuation. Greatest seasonal heave occurs in semiarid climates that have pronounced, short wet periods	Holland and Lawrence (1980)
2.2 Groundwater	Shallow water tables provide a source of moisture and fluctuating water tables contribute to moisture	
2.3 Drainage and manmade water sources	Surface drainage features, such as ponding around a poorly graded house foundation, provide sources of water at the surface; leaky plumbing can give the soil access to water at greater depth	Krazynski (1980); Donaldson (1965)
2.4 Vegetation	Trees, shrubs, and grasses deplete moisture from the soil through transpiration, and cause the soil to be differentially wetted in areas of varying vegetation	Buckley (1974)

2.5 Permeability	Soils with higher permeabilities, particularly due to fissures and cracks in the field soil mass, allow faster migration of water and promote faster rates of swell	Wise and Hudson (1971); De Bruijn (1965)
2.6 Temperature	Increasing temperatures cause moisture to diffuse to cooler areas beneath pavements and buildings	Johnson and Stroman (1976); Hamilton (1969)
3. Stress conditions 3.1 Stress history	An overconsolidated soil is more expansive than the same soil at the same void ratio, but normally consolidated. Swell pressures can increase on aging of compacted clays, but amount of swell under light loading has been shown to be unaffected by aging. Repeated wetting and drying tend to reduce swell in laboratory samples, but after a certain number of wetting–drying cycles, swell is unaffected	Mitchell (1976); Kassiff and Baker (1971)
3.2 In situ conditions	The initial stress state in a soil must be estimated in order to evaluate the probable consequences of loading the soil mass and/or altering the moisture environment therein. The initial effective stresses can be roughly determined through sampling and testing in a laboratory, or by making in situ measurements and observations	
3.3 Loading	Magnitude of surcharge load determines the amount of volume change that will occur for a given moisture content and density. An externally applied load acts to balance interparticle repulsive forces and reduces swell	Holtz (1959)
3.4 Soil profile	The thickness and location of potentially expansive layers in the profile considerably influence potential movement. Gratest movement will occur in profiles that have expansive clays extending from the surface to depths below the active zone. Less movement will occur if expansive soil is overlain by nonexpansive material or overlies bedrock at a shallow depth	Holland and Lawrence (1980)

2.2.1   Soil Characteristics

Soil characteristics may be considered either as microscale or macroscale factors. Microscale factors include the mineralogical and chemical properties of the soil. Macroscale factors include the engineering properties of the soil, which in turn are dictated by the microscale factors.

2.2.1.1   Microscale Factors: Clay Mineralogy and Soil Water Chemistry

Clay minerals of different types typically exhibit different swelling potentials because of variations in the electrical field associated with each mineral. The swelling capacity of an entire soil mass depends on the amount and type of clay minerals in the soil, the arrangement and specific surface area of the clay particles, and the chemistry of the soil water surrounding those particles.

Most clay minerals have orderly arrangements of atoms that form characteristic crystal lattices. The crystal lattice is the regularly repeated three-dimensional arrangement of atoms or ions in a crystal. An important characteristic of clay minerals is the small size of their crystals. Typical thicknesses can be as small as 15 Å and lateral dimensions are on the order of microns. This makes visual or standard microscopic examination impossible. X-Ray diffraction methods and electron microscopy have been used for identification on the basis of crystal structure. Different clay minerals also may be identified using chemical analyses. However, because each group has somewhat similar engineering properties, structural groupings are more convenient.

Three important structural groups of clay minerals are described for engineering purposes as follows:

- Kaolinite group—generally nonexpansive.
- Mica-like group—includes illites and vermiculites, which can be expansive, but generally do not pose significant problems.
- Smectite group—includes montmorillonites, which are highly expansive and are the most troublesome clay minerals.

All three of these groups have layered crystal structures. The mineralogical distinction is based on the physical arrangement of the different layers and the manner by which the individual structural units are bonded together. X-Ray diffraction provides a measure of the characteristic basal spacing, which describes the thickness and spacing of the individual bonded layers. Particle features and engineering properties of the important clay minerals are summarized in Table 2.3.

Soil water chemistry is important in relation to potential swell magnitude. Salt cations, such as sodium, calcium, magnesium, and potassium, are dissolved in the soil water and are adsorbed on the clay surfaces as exchangeable cations to balance the negative electrical surface charges. Hydration of these cations and adsorptive forces exerted by the clay crystals themselves can cause the accumulation of a large amount of water between the clay particles.

TABLE 2.3. Characteristics of some clay minerals

Mineral Group	Basal Spacing (Å)	Particle Features	Interlayer Bonding	Specific Surface (m²/g)	Atterberg Limits[a]			Activity[b] (PI/% Clay)
					LL (%)	PL (%)	SI (%)	
Kaolinites	14.4	Thick, stiff 6-sided flakes 0.1 to 4 × 0.05 to 2 μm	Strong hydrogen bonds	10–20	30–100	25–40	25–29	0.38
Illites	10	Thin, stacked plates 0.003 to 0.1 × 1.0 to 10 μm	Strong potassium bonds	65–100	60–120	35–60	15–17	0.9
Montmorillonites	9.6	Thin, filmy, flakes >10 Å × 1.0 to 10 μm	Very weak van der Waals bonds	700–840	100–900	50–100	8.5–15	7.2

[a]LL, PL, SL, liquid, plastic, and shrinkage limits, respectively.
[b]From Skempton (1953).
Summarized from Mitchell (1976).

In air dry soils, salt cations are held close to the clay crystal surfaces by strong electrostatic forces. As water becomes available, cation hydration energies are sufficiently large to overcome interparticle attraction forces. Thus, initially desiccated and densely packed particles are forced apart as adsorbed cations hydrate and become enlarged on the addition of water. When sufficient water is present, adsorbed cations are no longer held so tightly by the clay surfaces. Electrostatic attraction forces are counteracted by the tendency for ions to diffuse toward the more dilute bulk solution away from the particle surfaces. The electrostatic attraction forces counter the effects of diffusion to some extent resulting in a greater concentration of cations near the particle surface. The negatively charged clay particle surface and the concentration of positive ions in solution adjacent to the particle form what is referred to as a *diffuse double layer* or DDL (Bohn et al., 1985).

Overlapping DDLs betwee clay particles generate interparticle repulsive forces, or microscale "swelling pressures." Interaction of the DDL and, hence, swelling potential, increases as the thickness of the DDL increases. The thickness of the DDL is controlled by many variables including the concentration and valence of the cations in the soil water. In general, a thicker DDl and greater swelling are associated with lower cation concentrations and/or the presence of cations with low valence (Mitchell, 1976). Thus, for the same soil mineralogy, more swelling would occur in a sample having exchangeable sodium (Na^+) cations than in a sample with calcium (Ca^{2+} or magnesium (Mg^{2+}) cations. Also, leaching of salt from the clay pore fluid will enhance swelling potential.

2.2.1.2 Macroscale Factors: Plasticity and Density

Macroscale soil properties reflect the microscale nature of the soil. Because they are more conveniently measured in engineering work than microscale factors, macro-scale characteristics are primary indicators of swelling behavior. Commonly determined properties such as soil plasticity and density can provide a great deal of insight regarding the expansive potential of soils.

Soil consistency, as quantified by the Atterberg limits, is the most widely used indicator of expansive potential. Most expansive soils can exist in a plastic condition over a wide range of moisture contents. This behavior results from the capacity of expansive clay minerals to contain large amounts of water between particles and yet retain a coherent structure through the interparticle electrical forces. Soil plasticity is influenced by the same microscale factors that control swell potential and provides, therefore, a useful indicator of swell potential.

The electrical force fields between particles are related to the interparticle spacing. Hence the dry density and physical arrangement of particles (i.e., fabric) will affect the swell potential. Increased soil density through compaction or natural depositional history leads to greater amounts of swell and higher swell pressure.

2.2.2 Environmental Conditions

The potential for a soil to imbibe or expel water will depend on the water content relative to the water deficiency of the soil. Initial moisture content influences the shrink–swell potential relative to possible limits, or ranges, in moisture content.

Moisture content alone is not a good indicator or predictor of shrink–swell potential. Instead, the moisture content relative to limiting moisture contents such as the plastic limit and shrinkage limit must be known. Water content changes below the shrinkage limit produce little or no change in volume. There are indications that as a soil imbibes water, little volume change occurs at water contents above the plastic limit. The availability of water to an expansive soil profile is influenced by many environmental and man-made factors, as indicated in Table 2.2. Generally, the upper few meters of the profile are subjected to the widest ranges of potential moisture variation. Also, overburden stress is low and the soil is not restrained against movement at shallow depths. This upper stratum of the profile therefore exhibits the major part of the shrinking and swelling and is referred to as the *active zone*.

Moisture variation in the active zone of a natural soil profile is affected by climatic cycles. Boundary moisture conditions are changed significantly at the surface simply by placing a moisture barrier on the soil such as a building floor slab or pavement. The moisture barrier eliminates evapotranspiration of water from the surface, which, in turn, changes the moisture profile in the subsoils. The active zone is discussed in more detail in Section 2.2.4.

Other obvious and direct causes of moisture variation result from altered drainage conditions or man-made sources of water, such as irrigation or leaky plumbing. The latter factors are difficult to quantify but must be controlled to the extent possible for each situation. For example, proper drainage and attention to landscaping are simple means of minimizing moisture fluctuations near structures.

2.2.3 State of Stress

Volume change is directly related to change in the state of stress in the soil. A reduction in total stress due to excavation of overlying material will result in rebound and heave of the surface. Heave in unsaturated soils is accompanied by imbibition of water and is time dependent.

In saturated soils the state of stress affecting soil behavior is defined in terms of effective stress

$$\sigma' = (\sigma - u_w) \tag{2.1}$$

where σ' = effective stress
$\quad \sigma$ = total stress
$\quad u_w$ = pore water pressure
The three-dimensional effective stress can be written as

$$\sigma' = \begin{bmatrix} (\sigma_1 - u_w) & 0 & 0 \\ 0 & (\sigma_2 - u_w) & 0 \\ 0 & 0 & (\sigma_3 - u_w) \end{bmatrix} \tag{2.2a}$$

where σ_1, σ_2, and σ_3 are the principal stresses. In directions other than principal directions the effective stress would be

$$
\sigma' = \begin{bmatrix} (\sigma_x - u_w) & \tau_{xy} & \tau_{xz} \\ \tau_{yx} & (\sigma_y - u_w) & \tau_{yz} \\ \tau_{zx} & \tau_{zy} & (\sigma_z - u_w) \end{bmatrix} \tag{2.2b}
$$

where σ_x, σ_y, and σ_z are the normal stresses in the x, y, and z directions and the off-diagonal terms (e.g., τ_{xy}) are the shear stresses.

The stress state variable in this case takes into account the interaction between two phases, water and solids. A third phase, air, must also be considered for unsaturated soils. The definition of the effective state of stress for unsaturated soil must include the pressure in the pore air and the interaction between pore air and pore water. Fredlund and Morgenstern (1977) showed that the state of stress in an unsaturated soil can be defined by two independent stress state variables. The state of stress would be defined by two independent stress tensors

$$
\sigma' = \begin{bmatrix} (\sigma_1 - u_a) & 0 & 0 \\ 0 & (\sigma_2 - u_a) & 0 \\ 0 & 0 & (\sigma_3 - u_a) \end{bmatrix} \tag{2.3a}
$$

$$
\mu = \begin{bmatrix} (u_a - u_w) & 0 & 0 \\ 0 & (u_a - u_w) & 0 \\ 0 & 0 & (u_a - u_w) \end{bmatrix} \tag{2.3b}
$$

where u_a is the pore air pressure and μ is the capillary tension in the pore water. The term μ is frequently referred to as matric suction. Matric suction will be discussed in detail at a later point.

Alternatively, the stress tensors could be expressed as

$$
\sigma' = \begin{bmatrix} (\sigma_1 - u_w) & 0 & 0 \\ 0 & (\sigma_2 - u_w) & 0 \\ 0 & 0 & (\sigma_3 - u_w) \end{bmatrix} \tag{2.4a}
$$

$$
\mu = \begin{bmatrix} (u_a - u_w) & 0 & 0 \\ 0 & (u_a - u_w) & 0 \\ 0 & 0 & (u_a - u_w) \end{bmatrix} \tag{2.4b}
$$

For directions other than the principal directions, the off-diagonal terms in Eq. (2.4a) would be the shear stresses as in Eq. (2.2b). Because fluid pressure is isotropic, the off-diagonal terms in Eq. (2.4b) will always be zero.

The state of stress could also be defined by specifying only Eqs. (2.3a) and (2.4a). The use of these two tensors, however, is not usually convenient.

Thus, the state of effective stress can be defined by any combination of two of three stress state variables. Only two stress tensors are independent of each other. If any two are defined, the third one can be defined from them.

Constitutive relationships can be defined in terms of these independent stress state variables. For example, one-dimensional volume change can be described by the equation

$$\Delta e = C_{t1} \, \Delta \log(\sigma - u_a) + C_{ml} \, \Delta \log(u_a - u_w) \qquad (2.5a)$$

or

$$\Delta e = C_{t2} \, \Delta \log(\sigma - u_w) + C_{m2} \, \Delta \log(u_a - u_w) \qquad (2.5b)$$

Two constitutive parameters are necessary in these equations, because two stress state variables are specified, as opposed to only one for saturated soils. This emphasizes the independent nature of the two stress state variables and the fact that they must be considered independently. Geotechnical engineers are accustomed to defining state of stress for saturated soils in terms of only one stress state variable, $(\sigma - u_w)$. It is necessary to consider two stress state variables for unsaturated expansive soil because the behavior is also affected by soil suction.

There exists a unique relationship for all soils between matric suction and water content. This relationship also depends on soil density and structure. This will be discused in detail in Chapter 4.

Changes in the soil moisture environment that result in stress changes must be considered in addition to those changes caused by the presence of structures. For example, vertical loads imposed by structures will reduce expansion, but the presence of the structure may also cause increases in moisture content that can cause swelling. It is evident, therefore, that the site investigation must include the ability to predict changes not only in the geostatic stresses, but also in environmental factors that will influence the soil suction.

EXAMPLE 1

Given.

An element of soil at a depth of 2 ft below the ground surface. The water table is at a depth of 5 ft below the ground surface. The pore water is continuous from the water table to the ground surface. The total unit weight of the soil is $\gamma_t = 110$ pcf. Assume that the *total* horizontal stress is half of the *total* vertical stress.

Find.

a. The matric suction at the soil element.
b. Write the stress tensors that describe the state of stress at the soil element.

Solution.

a.

$$u_a = 0 \text{ (atmospheric)}$$

$$u_w = \gamma_w z \ (z = \text{depth below water table})$$

$$= 62.4 \text{ pcf} \times (2 - 5) \text{ ft}$$

$$= -187.2 \text{ psf}$$

$$(u_a - u_w) = [0 - (-187.2)] = 187.2 \text{ psf}$$

b. Write stress tensors in terms of $(\sigma - u_a)$ and $(u_a - u_w)$.

$$\sigma_v = 2 \times 110 = 220 \text{ psf}$$

$$\sigma_h = 0.5 \times 220 = 110 \text{ psf}$$

From above,

$$u_a = 0$$

$$u_w = -187.2 \text{ psf}$$

Therefore,

$$(\sigma_v - u_a) = 220 - 0 = 220 \text{ psf}$$

$$(\sigma_h - u_a) = 110 - 0 = 110 \text{ psf}$$

Also

$$(\sigma_v - u_w) = 220 - (-187.2) = 407.2 \text{ psf}$$

$$(\sigma_h - u_w) = 110 - (-187.2) = 297.2 \text{ psf}$$

The stress tensors are,

$$(\sigma - u_a) = \begin{bmatrix} (\sigma_v - u_a) & 0 & 0 \\ 0 & (\sigma_h - u_a) & 0 \\ 0 & 0 & (\sigma_h - u_a) \end{bmatrix} = \begin{bmatrix} 220 & 0 & 0 \\ 0 & 110 & 0 \\ 0 & 0 & 110 \end{bmatrix}$$

and

$$(u_a - u_w) = \begin{bmatrix} (u_a - u_w) & 0 & 0 \\ 0 & (u_a - u_w) & 0 \\ 0 & 0 & (u_a - u_w) \end{bmatrix} = \begin{bmatrix} 187.2 & 0 & 0 \\ 0 & 187.2 & 0 \\ 0 & 0 & 187.2 \end{bmatrix}$$

Either of the above could be replaced by

$$(\sigma - u_w) = \begin{bmatrix} (\sigma_v - u_w) & 0 & 0 \\ 0 & (\sigma_h - u_w) & 0 \\ 0 & 0 & (\sigma_h - u_w) \end{bmatrix} = \begin{bmatrix} 407.2 & 0 & 0 \\ 0 & 297.2 & 0 \\ 0 & 0 & 297.2 \end{bmatrix}$$

2.2.4 Active Zone

Expansive soil problems typically arise as a result of an increase in water content in the upper few meters. There have been instances of deep-seated heave, but these are rare. The water contents in these upper few meters are influenced by climatic environmental factors. This zone is generally termed either the *zone of seasonal fluctuation* or the *active zone*.

Figure 2.2 depicts the active zone. The water content distribution in the active zone is different than what would exist under hydrostatic equilibrium conditions as

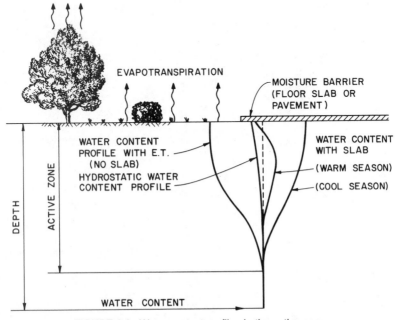

FIGURE 2.2. Water content profiles in the active zone.

a result of climatic factors and evapotranspiration. Hydrostatic conditions will dictate a negative pore water pressure (soil suction) above the groundwater table that varies linearly with height above the groundwater level. The degree of saturation will be less than or equal to 100% in this zone depending on the water retention characteristics of the soil. The hydrostatic water content profile would be as shown in Figure 2.2. Due to evapotranspiration from the surface the water content in arid or semiarid regions would be lower (drier) than the hydrostatic water content profile. That curve is also shown in Figure 2.2.

If excess water is added at the surface or if evapotranspiration is eliminated, the water content will increase and heave will occur. In addition, temperature gradients can influence the migration of water. In Figure 2.2 the water content profiles for cool seasons and warm seasons are shown. The effect of temperature gradients would depend on the temperature regime in the soils. Beneath slabs in houses the effect of temperature will normally be less pronounced near the center of the slab than near the edge. For uncovered slabs in cold climates, the temperature gradient can be appreciable even under the center of the slab. The concept of the active zone will be discussed more fully at several points later on.

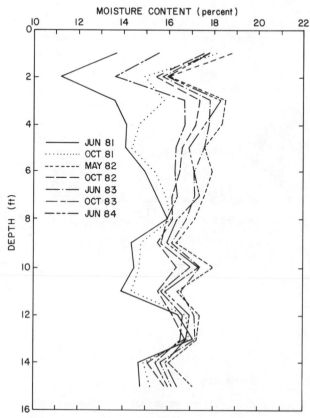

FIGURE 2.3. Seasonal water content profiles (Hamberg, 1985).

The site investigation should provide information that will allow determination of the active zone. In arid or semiarid locations, water contents will typically increase with depth until a point at which the water content becomes constant with depth. Figure 2.3 shows the water content profile for a site in Colorado (Hamberg, 1985). This profile was developed by taking moisture readings over several seasons after a simulated floor slab had been placed on the ground surface. Seasonal moisture fluctuations were found to be small below a depth of 5.5 m (13 ft), indicating that the active zone is about that deep at this site.

By developing similar plots of water content as a function of depth over several wet and dry seasons, the active zone depth may be estimated for any site. The depth at which the water content becomes nearly constant with depth should define the active zone depth. In cases where the soil is not uniform with depth, or if several strata exist, the differences in soil type can be compensated for by plotting either water content divided by plasticity index (w/PI) or liquidity index [$(LL - w)/PI$], rather than water content.

Analytical problems generally are associated with underestimation of the depth of the active zone. Along the Front Range of Colorado, the actives zone appears to be generally about 15 to 20 ft deep. Estimates of active zone depths less than 10 ft should be considered suspect.

EXAMPLE 2

Given.

The soil profile and data below.

Soil	Depth (ft)	Plastic Limit	PI (%)	Water Content (%)
0 to 6 ft	2.5	22	28	11
Clay, stiff, moist, mottled gray, and olive green	5.0	22	28	13
6 to 10 ft Clayey silt, dense, dry to moist, tan	8.0	12	8	5
10 to 12 ft Silty sand, dry, brown	11.5	np	np	3
12 to 35 ft	13.5	26	36	16
Clay, stiff, moist,	16.5	26	36	19
dark brown	20.0	26	36	28
	25.0	26	36	29
	30.0	26	36	28

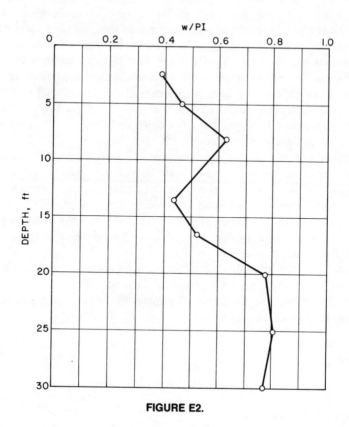

FIGURE E2.

Find.

Estimate the depth of the active zone.

Solution.

Plot w/PI as shown in Figure E2. At depths greater than about 20 ft, the values of w/PI become approximately constant. The depth of the active zone is probably about 20 ft.

2.3 SITE EXPLORATION

2.3.1 General Considerations of Exploration

The geotechnical site investigation is a unique process of material characterization. It is not uncommon for the site investigation to include material sampling and testing of the foundation soil at a much lower frequency per unit volume than the structural concrete, although most soil profiles are far more complex and variable. The geotechnical investigator is charged with the task of defining and characterizing large volumes of material using widely spaced, point sources of data.

Because the soil data are incomplete, the geotechnical engineer must avoid the tendency to overanalyze an insufficient quantity of data. Precise analysis of incomplete data can lead to inaccurate design assumptions. If a choice must be made between more accurate definition of the soil profile and precise determination of soil properties by sophisticated laboratory techniques, the profile should receive primary attention. The following statements, from an ASCE Specialty Workshop on Site Characterization and Exploration, are worthy of note (Dowding, 1979):

- New *gigits* (unrealistically presumed, forthcoming methods for filling in gaps) will not substitute for adequate information.
- Exploration must be conducted by trained personnel who are familiar with the reasons for exploration.
- Test results and their analysis are only as valid as the quality and extent of the field investigation on which they are based.
- Overrefinement of analysis does not lead to improved design.

Flexibility in the exploration program is essential, especially pending the outcome of the preliminary investigation. The preliminary investigation forms the basis for planning the costlier and more specialized detailed investigation that follows. If a program is too inflexible, and preconceives site evaluation requirements, it may leave out pertinent information but include unnecessary information (Osterberg, 1979).

The required intensity of site exploration will depend on the size and nature of the project, the geologic complexity, and the stage of progress in the investigation. The primary objective in any program of site characterization should be to acquire a sufficient amount of information, in such a way that the geotechnical concerns associated with a particular project can be satisfactorily resolved (Mitchell, 1979).

Although there will be special concerns associated with characterizing sites having expansive soils, especially in the evaluation of heave potential, the objectives of the investigation will follow the general outline of any other geotechnical exploration program. The purposes of the geotechnical site investigation and means of accomplishing those purposes are summarized in Table 2.4.

2.3.2   Program of Exploration

Soil exploration methodologies are usually step-by-step processes that develop as information accumulates. The staged procedure involves the following steps:

1. Reconnaissance—Review available information and perform a surface reconnaissance.

TABLE 2.4. The geotechnical site investigation

Purpose	Methods
Define the geometry of relatively homogeneous zones of the soil and rock profile	Maps, field, and aerial surveys for surficial geology and topography
	Geophysical methods, auger and sample borings for subsurface profile
Characterize the soil conditions within each homogeneous zone	Visual field observations during borehole logging
	Field testing methods
	Laboratory testing
Determine the ground moisture conditions, and potential fluctuations thereof	Locate and monitor the groundwater table using observation wells, piezometers, or geophysical techniques
	Locate natural and man-made drainage features and extent of floodplains using maps and field observations
	Assess climate and environmental conditions such as amount and variation of precipitation and type of vegetation, using Weather Service records, field observations, etc.
Obtain site history information regarding performance of other structures. Removal of trees and shrubs. Existence of earlier structures, ponds, and depressions, or water-courses. Previous changes in groundwater conditions, etc.	Interviews with local engineers and others
	Field observations
	Local records
	Technical literature

2. Preliminary Investigation—Conduct detailed surface mapping, preliminary borings, and initial laboratory testing and analysis for soil identification and classification.
3. Detailed Investigation—Conduct soil borings for recovery of specialized samples for testing and analysis, conduct specialized field tests, and partial excavation.

Results from each stage should be integrated and evaluated for their impact on design requirements and construction procedures.

A reconnaissance stage should always precede any drilling program. The investigator will often be able to find reports and technical articles such as Soil Conservation Service survey reports or previous soil investigations. Other useful information that can be obtained prior to the field investigation includes observations of the behavior of other structures in the area, indications of slope instability, weather and rainfall data, maps and reports from state and federal geological surveys, and agricultural soils surveys. This information can be very useful for planning the areal distribution and intensity of the initial drilling program.

Some areas have been mapped for expansive soils, through correlations with geologic and agricultural soils maps (see Krohn and Slossan, 1980; Hart, 1974; and Snethen et al., 1975, for example). Figure 2.4 shows a map of the general distribution of expansive materials in the United States. Geologic and soils maps are particularly useful in transportation work where general rock and soil features must be evaluated over large areas.

A good knowledge of local geology and site location relative to other problem areas is very useful in the identification of potential problems. Awareness of those particular formations that exhibit high swelling potential is necessary for the local practicing geotechnical engineer. In northern Florida, for example, soils from the Hawthorne formation may exhibit shrink–swell characteristics. Reasonable correlation has been established in the Gainesville area between geographic locations of expansive foundation soil problems and certain elevations in the Hawthorne formation. As another example, the Pierre formation along the Front Range in Colorado contains expansive clay shale. Geologic maps showing its location near the surface are very useful in identifying expansive soil potential.

Field observations made during the reconnaissance and preliminary investigation phases can provide valuable data, and can be obtained easily, even by relatively inexperienced professionals with some on-the-job training (Krazynski, 1976). The field characteristics to look for will be somewhat localized. Some examples of important observations that can be obtained by the field engineer are as follows.

1. Soil characteristics:
   - Spacing and width of wide or deep shrinkage cracks.
   - High dry strength and low wet strength (indicates high plasticity).
   - Stickiness and low trafficability when wet.
   - Scraped or cut surfaces have a glazed or shiny appearance, like soap.

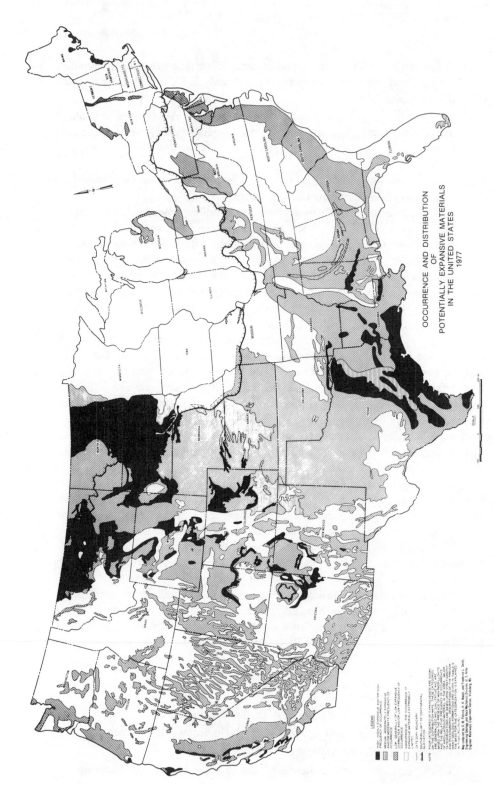

FIGURE 2.4. Distribution of potentially expansive soils in the United States (U.S. Army Engineer Waterways Experiment Station, 1977).

2. Geology and topography (ground or aerial surveys):
   - Hummocky topography, known as *gilgai* formation.
   - Evidence of low permeability indicated by surface drainage and infiltration features.
3. Environmental conditions:
   - Vegetation type. For example, in central Texas mesquite will usually grow in expansive soils but oak will not. Pine forests in Houston usually indicate safe soils. Vegetation should also be noted for potential design problems. Shrubs and trees with spreading roots, such as elms and cottonwoods, can cause local soil desiccation, which may lead to differential soil movements under structures situated near them.
   - Climate. Arid and semiarid areas are particular trouble spots because of large variations in rainfall and temperature.

Climate rating systems are available, the most popular of which is the Thornthwaite Moisture Index (TMI) (Thornthwaite, 1948). The TMI categorizes climate by balancing the rainfall (R), potential evapotranspiration (PE), and soil water holding capacity (HC). Negative TMI values indicate dry climates. TMI values between +20 and −20 indicate areas most prone to having problems (O'Neill and Poormoayed, 1980). Figure 2.5 shows the TMI distribution in the United States.

The Building Research Advisory Board (BRAB, 1968) published a climatic rating map, shown in Figure 2.6, for use in foundation slab design (see Figure 2.6). Climatic ratings (C_w) were determined from U.S. Weather Bureau information concerning yearly annual precipitation, number of times precipitation occurs, duration of each occurrence, and amount of precipitation at each occurrence. The larger the numerical rating, the greater the abundance of moisture.

2.3.3 Drilling and Sampling

The preliminary drilling program should be planned to emphasize sampling in areas where the reconnaissance indicates that problems may exist. Depending on the results of that program a more detailed drilling program may be designed or the preliminary program may suffice. If a more detailed program is undertaken, the value of the preliminary program is to provide initial data to aid in locating areas of sampling, frequency of sampling, types of in situ measurements to be made, and definition of other elements of the detailed program.

Although sampling methods utilized for expansive soils are generally the same as those used for conventional soil sampling care must be taken to avoid disturbance such as moisture content increase from drilling fluids. Disturbed samples from auger cuttings or large borings may be used to log the subsurface profile and for classification tests such as Atterberg limits, specific gravity, and grain size distribution. Design testing requires high quality, undisturbed samples.

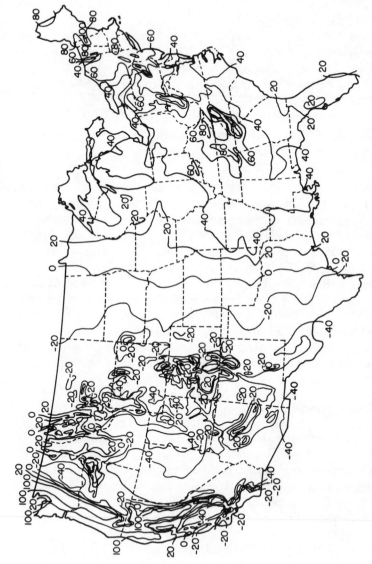

FIGURE 2.5. Thornthwaite Moisture Index distribution in the United States (Thornthwaite, 1948).

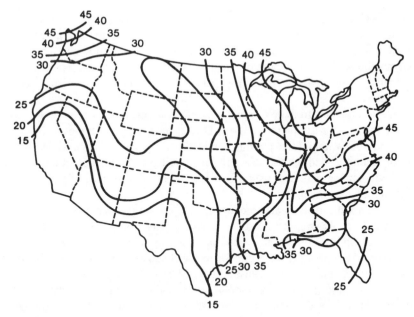

FIGURE 2.6. Climatic ratings (C_w) for the United States (BRAB, 1968).

2.3.3.1 Sampling for Purposes of Classification

Auge samples can provide valuable disturbed material for the preliminary investigation. Single flight power augers, which can be mounted on trucks or tracked vehicles, can be used to cover large areas in difficult terrain. Samples are taken from the cutter heads or from special sampler attachments. Single flight augers are generally limited to depths of about 20 to 30 ft (6 to 9 m).

Continuous flight augers can sample to depths of 100 ft (30 m) or more. However, the original depth location of the soil carried to the top of the hole is uncertain. This soil may also be mixed with other soil along the length of the auger. If accurate depth location is needed samples should be taken from the cutter head, and it should be pulled up in intervals from known elevations.

Hollow stem augers, which are continuous flight augers with hollow centers, provide a means of obtaining a clean open hole for sampling undisturbed specimens.

Collecting samples from auger cuttings below the water table is very difficult and generally not recommended (Acker, 1974).

2.3.3.2 Undisturbed Sampling for Testing

A large variety of undisturbed sampling techniques have been developed in soils engineering. The choice of which type of sampler to use is governed largely by the stiffness of the soil being sampled. Expansive soils can vary from medium firm or firm materials, such as the Prairie Terrace formations of Louisiana, to very stiff materials, such as the Pierre, Mancos, and Bearpaw shales of the Northern and Central Plains of the United States and Canada.

Thin walled tube samplers, or Shelby tubes, are used in areas where the soils are sufficiently soft to allow the sampler to be pushed into the soil properly. The objective in push tube sampling is to advance the thin wall tube into the soil with one smooth stroke. Driving push tube samplers with weights is unacceptable practice in expansive soil investigations. The ASTM Specification, D-1587-67, specifies thin wall sampler sizes.

If the soil is too hard to be sampled using a push tube type sampler, either driven tube or rotary core samplers are used. Driven tube samplers typically have thick walls and do not, therefore, provide "undisturbed" samples. Thick walled samples are not recommended for use in testing for swell prediction. Poor quality, highly disturbed samples may seriously affect laboratory results and impair the reliable prediction of movement.

In spite of the disturbance produced, a type of drive sampler known as the California drive tube sampler is used extensively in the Front Range area of Colorado and Denver, in the Los Angeles area, and possibly other areas. Figure 2.7 shows a drawing of this sampler. It is similar to the Standard Penetration Test split tube sampler but includes brass or plastic liners. Although the liners may reduce side wall friction, these samplers are not thin walled and the samples must be considered disturbed. Driving of the relatively thick walled sampler generally causes a substantial increase in density of the sample. The measured swelling characteristics, therefore, overestimate the percent swell and the swelling pressure. It is the authors' opinions that because of this built-in conservatism this method of sampling has not been rejected by the geotechnical engineering community. It is economical, it provides samples that can be tested, and the error is on the conservative side.

More sophisticated, double-barrel rotary core samplers such as the Denison, Pitcher, or WES types are used to sample undisturbed material in stiffer formations. These samplers are more expensive than push tube types, but are more suitable for hard soil and rock and soils containing gravel. Rotary core samplers consist of an inner and outer barrel, with the cutting bit on the outside and above the inner, protruding sampler barrel. The inner barrel is a full swivel type mounted on antifriction bearings and remains stationary during drilling. The core barrel is forced downward with gradually increased pressure, collecting the sample into a thin walled liner.

2.3.3.3 Sample Disturbance

Sample disturbance is a problem in any soil testing program, but it is a particularly important consideration when sampling and testing expansive soils. Sample disturbance is minimized by using thin walled sampling containers and by reducing the frictional resistance between the sample tube and the soil. Lubrication can be provided by spraying the tube with silicon or Teflon spray. Care must be taken to avoid the use of lubricants that might react with the soil and change its properties. Polishing, or plating the inner surface of the sampler or the use of foil lined samplers may be effective in reducing disturbance. If possible, the soil samples should be kept in carefully sealed sampling tubes until just before testing to avoid disturbance due to stress relief.

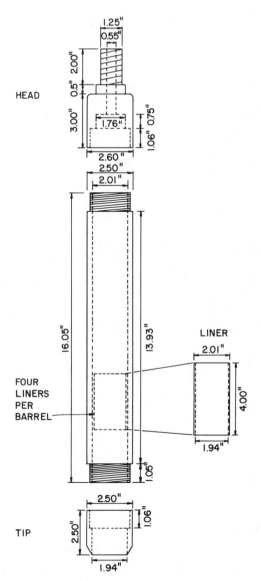

FIGURE 2.7. California drive tube sampler.

2.3.3.4 Sampling Frequency

Soil sampling programs for different applications or projects will involve different strategies. For example, a sampling program for a building will involve more closely spaced samples at greater depths than that for a highway or airfield project. Nevertheless, in most applications it is common practice to take a greater number of samples and to somewhat deeper depths in expansive soil areas.

In foundation investigations, the number of borings made per site varies considerably. Typically, two to six borings are made for a single structure. If the location of a small structure is well defined, four or less holes are commonly drilled. It is recommended that holes be drilled at all corners of the structure or at least at both ends of the structure. In residential subdivisions, recommended practice is to take at least one boring per house lot. Two borings per lot is not unreasonable if rough grading is extensive and a large number of cut and fill lots are present.

Sampling intervals should be frequent in each borehole. Krazynski (1979) recommended sample recovery at 1 ft (30 cm) intervals in the uppermost 5 ft (1.5 m). Common practice is to take two samples in the uppermost 5 ft and one at every 5 ft interval down to depths of 20 ft (6 m). Sampling intervals may be increased for greater depths.

Samples should be taken down to depths below the depth of seasonal moisture fluctuation or to depths below which the loads applied by the structure have an insignificant effect, whichever is greater. Holes should be logged carefully in the field to avoid confusion later during testing and analysis.

Development of a drilling program for highway investigations typically involves extensive exploratory information and engineering judgment regarding the geological overview during the reconnaissance and preliminary phases. The strategy in subgrade investigation for pavement design is based on analysis of the soils within a design unit. These units are delineated initially on the basis of maps, geology, and climate. Design unit boundaries are finalized after laboratory and field testing. Design values for parameters such as the California Bearing Ratio (CBR) are then selected based on testing of the controlling soil within each design unit (Yoder and Witczak, 1975).

Sampling practice can vary considerably from one location to another. Typical practice within state highway departments in the United States for subgrade evaluation involves AASHTO and Unified Soil Classification testing only, except for special projects.

General considerations and guidelines for proper sampling, handling, and shipping of soil samples are described in Acker (1974), ASTM (1971), Hvorslev (1948), and U.S. Bureau of Reclamation (1974).

2.3.4 Field Testing

In situ measurements not only provide valuable data needed to assess soil conditions for characterization of the profile, but also allow the quantitative assessment of soil properties in place. They have the distinct advantage of allowing measurements under the field environmental conditions and with a minimum amount of disturbance due to sampling.

Some of the following field tests are recommended in the characterization or evaluation of expansive soil sites:

- Soil suction measurements using in situ thermocouple psychrometers and tensiometers, or filter paper testing methods.
- In situ density and moisture tests.

- Surface and subsurface heave and settlement monitoring.
- Piezometers or observation wells for locating and monitoring the water table.
- Penetration resistance.
- Pressuremeter and dilatometer tests.
- Geophysical methods.

2.3.4.1 In Situ Soil Suction Measurement

It was shown in Section 2.2 that soil suction is of particular importance because it is one of the stress state variables controlling the behavior of the soil. Soil suction is expressed as a negative gage pressure, which can be measured directly by capillary principles at low suctions, or by potential energy (psychometric) techniques at higher suctions.

Typical units of suction are summarized in Table 2.5. It should be noted that a positive value of suction indicates a negative pore water pressure. The second column in Table 2.5 shows units in pF. The pF value is expressed without dimensions and is the logarithm to the base 10 of the capillary head, i.e., suction head, expressed in centimenters of water. This unit was used fairly frequently in the 1950s and 1960s. However, conventional units of stress (e.g., psi or kPa) or height of water (ft H_2O or m H_2O) are finding more widespread use.

In situ suction values may range from zero to over 15,000 psi (100,000 kPa). Suction values as high as 150,000 psi (10^6 kPa) have been reported (Olson and Langfelder, 1965). These high values include a predominant component of osmotic suction.

In situ suction determination techniques were developed primarily by soil scientists for agricultural purposes. At this point a brief discussion of different means of measuring suction in the field will be presented. A detailed discussion of soil suction and its measurement is given in Chapter 4.

Tensiometers. The tensiometer measures matric suction by capillary principles. Tensiometers consist of saturated, very finely grained porous ceramic cups. When buried in the soil and connected to a device capable of measuring negative pressure, the tensiometer will come to equilibrium pressure with the water in the soil pores. Application of these instruments is limited to engineering work. At suction values much above 7 to 8 psi (50 kPa), or about half an atmosphere, the water cavitates (air comes out of solution) and the continuous water column in the tensiometer breaks. Expansive soils frequently have suctions well above the limits of measurement by tensiometers. They are useful, however, for low suction values, for applications in wet soils, and for agricultural applications.

Heat Dissipation Sensor (Thermal Matric Potential). The heat conductivity in a material is a function of the water content of that material. By measuring the rate of heat dissipation, the water content of the material can be obtained. Instrumentation has been developed wherein a standardized porous ceramic probe is inserted into the soil and allowed to come into equilibrium with the soil. Water will flow into

TABLE 2.5. Relationship between various suction units used

Height of Water Column (cm)	pF	psi	kgf/cm²	kPa	Bars	Atmos.
1	0	0.0142	0.001	0.0981	0.00098	0.00096
10	1	0.1422	0.01	0.981	0.0098	0.0096
10^2	2	1.422	0.1	9.81	0.098	0.0968
10^3	3	14.22	1	98.1	0.981	0.968
10^4	4	142.2	10	981	9.81	9.68
10^5	5	1,422	10^2	9,810	98.1	96.8
10^6	6	14,220	10^3	98,100	981	968
10^7	7	10^4	10^4	981,000	9,810	9,680

After Fargher et al. (1979).

or out of the ceramic probe until the suction is the same in both the soil and the probe. After equilibrium is reached, a controlled heat flux is applied to the center of the porous probe, and the increase in temperature is measured over a fixed period of time. The change in temperature is indicative of the water content of the porous ceramic material. The suction of the probe, and hence the soil, is directly related to the water content.

Thermocouple Psychrometers. Spanner (1951) applied the meteorological principle of using wet bulb and dry bulb temperatures to determine relative humidity and, hence, soil suction in unsaturated soils. The relative vapor pressure at a point in a soil mass can be related to the total soil pressure at that point.

Thermocouple psychorometers (TCPs) are used to measure relative humidity in accordance with Peltier cooling and Seebeck thermoelectric effects. The devices are calibrated over aqueous salt solutions under controlled temperature conditions.

Details of calibration and use of thermocouple psychrometers were reported by McKeen (1981), Snethen (1979b), Riggle (1978), Meyn and White (1972), Slack (1975), Brown and Van Haveren (1972), and Campbell and Gardner (1971). More detailed discussion about the principles and calibration techniques for TCPs are presented in Chapter 4.

One of the greatest advantages of the instruments is that they provide a means of measuring high values of soil suction under in situ conditions. Some disadvantages of TCPs for field applications are as follows:

- Exposure to moist conditions for long periods of time will corrode the instruments (McKeen, 1980; Johnson and McAnear, 1974).
- Each instrument must be calibrated individually, and calibration checks should be performed periodically (Baker et. al., 1973).
- Low sensitivity at suctions below about 50 psi (300 kPa) make these instruments unusable in moist soils. Profiles that remain below this level of suction should not be instrumented. TCPs are satisfactory for use between about 50 and 150 psi (300 and 100 kPa) and are best at suction values above 150 psi (1000 kPa) McKeen, 1981).

Although routine field applications have not yet occurred, TCPs hold promise as potentially reliable and convenient in situ suction sensors.

Filter Paper Sensors. A wide range gravimetric method for moisture suction determination using filter paper disks was adopted for engineering purposes by McQueen and Miller (1968, 1974), McKeen (1980), and Johnson and Snethen (1978). Basically, calibrated filter papers are equilibrated with natural soil samples in a closed container for a period of time at a constant temperature. After equilibration, the filter papers are removed, weighed and dried for moisture content determination, and correlated with soil suction. Recent research has utilized other absorbent materials instead of filter paper. This can increase the sensitivity of the method over different ranges of suction (Sibley and Williams, 1990).

Filter paper calibration is performed in the suction range of interest. At high suctions, the disks are equilibrated over salt or acid solutions. Calibrations at suctions below about 150 psi (1000 kPa) are attained by using a pressure plate or pressure membrane apparatus to determine soil suction in standard soil samples. The filter paper may be calibrated against those soil samples. More detailed discussion of calibrations and weighing of filter papers are presented in Chapter 4.

The filter paper test is simple to perform and inexpensive. Field samples are immediately sealed on site in moisture jars along with filter papers. The jars are stored in simple insulated containers such as ice chests.

The filter paper test is unsuitable in areas of high humidity, and the 7-day equilibration period may be disadvantageous in some situations. However, the low cost and simplicity of the technique along with the fact that filter papers can be used to measure the entire range of soil suction are advantages that outweigh the problems of this method. Filter paper tests are especially useful for investigations over large areas, such as airfields and highways, because a large number of observations can be taken easily.

2.3.4.2   *In Situ Measurement of Stress and Soil Properties*

Nuclear Moisture Probes.   Nuclear methods of in situ moisture content determination have been used since the mid-1950s. In civil engineering practice, the commonly used device for in situ moisture determination is the neutron probe.

Neutron moisture equipment consists of a probe, containing a source of fast neutrons, suspended on a cable that attaches to a scaler and counting rate recorder. When the probe is inserted into a mass of soil, the fast neutrons are scattered and slowed by collisions with hydrogen nuclei. The concentration of slowed neutrons near the source is detected and monitored using the counting device. Counting rate is correlated with the amount of water per unit volume of soil. The volume of soil affecting the neutron count decreases with increasing water content.

Moisture probes are lowered into aluminum or polyethylene access tubes for determining the moisture content. Studies conducted over several wetting and drying periods provide valuable information about depths of seasonal moisture change.

Manufacturers of neutron moisture meters provide calibration curves, but a separate calibration for each site is recommended. On-site calibration is accomplished by taking gravimetric moisture content samples near the access tubes, or by compacting the soil in large drums at different water contents and field bulk densities.

Standard Penetration and Cone Penetration Tests.   The most commonly used in situ test is the dynamic standard penetration test (SPT). The SPT interpretation relies on empiricism. A discussion of theoretical aspects of the test is presented by Schmertmann (1976). The SPT is most useful in expansive soil investigations as a preliminary indicator of material variability in a soil profile, and in correlation with in situ density. Chen (1988) correlated SPT results, together with soil index properties, to swell potential of soils in the Rocky Mountain Region of the United States (see Table 3.3 in Chapter 3).

The static, or Dutch cone penetration (CPT) may also be used as an indicator of soil type. Static cone resistance, q_c, has been related to soil compressibility for many normally consolidated and soft soils. Most expansive soils are usually overconsolidated and these correlations are not valid for them. A more appropriate use of CPT data in stiffer clays is for correlation with the overconsolidation ratio (OCR) using a method developed by Schmertmann (1974, 1975). In this method the undrained shear strength, S_u, is estimated based on q_c at varying depths. Knowing the corresponding effective overburden stress, p_0', at each depth (assuming the water table has been located), the ratio S_u/p_0' can be computed. Relationships between S_u/p_0' and OCR for various clays can then be used to provide an estimate of the OCR.

Dilatometer. Marchetti (1975, 1980) developed a device known as the flat plate dilatometer. This device consists of a 95-mm-wide by 14-mm-thick stainless steel blade, with a circular expandable steel membrane on one side. The equipment is shown in Figure 2.8. The dilatometer is jacked into the ground using conventional penetrometer equipment. At specified intervals, jacking is stopped and the membrane is inflated against the surrounding soil. Two pressure readings are taken. The first reading is taken at the pressure required to just begin to move the membrane, and the second reading is taken at the pressure required to push the center of the membrane 1 mm into the soil. Each set of readings takes between 15 and 30 sec.

Dilatometer test results are used to compute a material index, I_D, a horizontal stress index, K_D, and a dilatometer modulus, E_D (Marchetti, 1980). The material index, I_D, has been correlated with predominant grain size, making this index useful in general soil classification. The horizontal stress index, K_D is related to the coefficient of earth pressure at rest, K_0, and stress history. The dilatometer modulus, E_D, is related to soil stiffness. The three parameters can be used together to provide a great deal of information about the material variability and in situ stress conditions in a soil profile. As use of the device becomes more widespread and standardized, more correlations will undoubtedly become available for all soil types, including expansive soils.

Pressuremeters. Pressuremeter tests are finding fairly widespread use in expansive soil investigations. The pressuremeter, shown in Figure 2.9, consists of a flexible cylindrical probe that is placed inside a drilled hole and expanded out against the sides of the hole. The volume changes are recorded with increasing pressure. Interpretation of test results utilize semiempirical relationships based on the theory for internal pressure expanding a cavity into an elastic half space.

The self-boring pressuremeter has particular advantages over pressuremeters that are placed in predrilled holes (Wroth and Hughes, 1983). This device consists of a hollow cylindrical tube with a pressuremeter membrane mounted flush with the outside. It can be drilled into the soil at a constant rate with the cuttings transported to the surface by circulating water or drilling mud. The self-boring instrument can be used to determine basic soil parameters related to in situ stress, strength, and compressibility (Davidson, 1983).

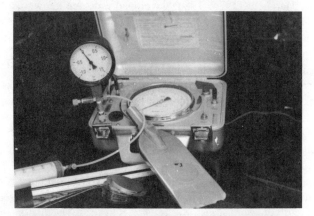

FIGURE 2.8. Flat plate dilatometer device and readout equipment.

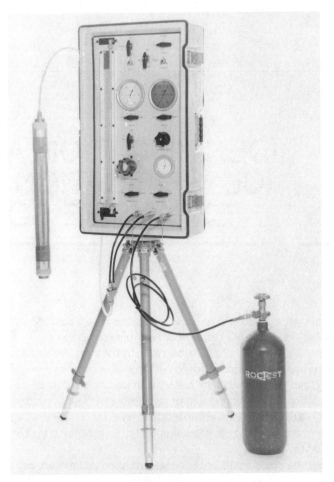

FIGURE 2.9. Pressuremeter (Courtesy of Roctest, Inc.).

In Situ Swell Pressure Device. A device for measuring in situ lateral swelling pressure has been described by Ofer (1988). The in situ swelling pressure probe consists of a hollow cylinder, having a reduced thickness in the center on which strain gages are mounted. Extending below the measuring module is a cutting cylinder that planes a smooth surface for the measuring module inside a slightly undersized, predrilled hole. Once the probe is positioned, two wetting rings, located above and below the measuring module, supply water to the soil surrounding the device. The swelling pressures are monitored using the strain gages attached to the thin portions of the cylinder.

This device has been used primarily for research purposes. It has been used in a limited number of applications for design purposes on projects in South Africa (Ofer, 1988).

3

IDENTIFICATION AND CLASSIFICATION OF EXPANSIVE SOILS

Early identification of expansive soils, during the reconnaissance and preliminary stages of a project, is essential to allow for appropriate sampling, testing, and design in later stages. Thus, the investigation must actually comprise two important phases. The first is the recognition and identification of the soil as expansive soil and the second is sampling and measurement of material properties to be used as the basis for design predictions. This chapter will discuss tests and classification procedures that have been used to identify expansion or shrink–swell potential. Testing of expansive soils to determine properties to be used in design predictions is treated as a separate topic in Chapter 4.

In engineering practice, the common identification schemes are based on standard classification results, such as grain size analysis and Atterberg limits. However, other tests used in identifying potentially swelling soils are available, and are used routinely by agricultural and geological survey laboratories. These methods should not be overlooked by the engineering community, because they provide valuable additional information about the mineralogical and chemical nature of soils. The tests are summarized in Table 3.1.

3.1 IDENTIFICATION TESTS

3.1.1 Engineering Classification Tests

Classification tests for soil index properties such as grain size distribution, clay content, and plasticity are the most widely used in practice for identifying and classifying expansive soils.

The Atterberg limits define moisture content boundaries between states of consistency of fine-grained soils. Figure 3.1 illustrates the concept originated by A.

Atterberg that a clay soil can exist in four distinct states of consistency depending on its water content. The water content at the boundaries between the different states are defined as the shrinkage, plastic and liquid limits.

Two useful indices may be computed from the Atterberg limits and the natural moisture content. These are the plasticity index (*PI*) and the liquidity index (*LI*), and are defined in Table 3.1. The *PI* is used extensively for classifying expansive soils and should always be determined during preliminary investigations.

The plasticity characteristics and volume change behavior of soils are directly related to the amount of colloidal sized particles in the soil. For engineering purposes the term *colloid* is used to describe a particle whose behavior is controlled by surface forces (i.e., electrostatic and adsorptive forces) rather than by gravitational forces. Colloid size is generally defined as being smaller than 0.001 mm. Most clay particles can be considered colloids by the engineering definition because of their irregular shapes and large surface areas.

Atterberg limits and clay content can be combined into a single parameter called Activity. This term was defined by Skempton (1953). The activity is defined as follows:

$$\text{Activity } (A_c) = \frac{\text{Plasticity index}}{\% \text{ by weight finer than } 2 \ \mu\text{m}} \tag{3.1}$$

Skempton suggested three classes of clays according to activity as *inactive*, for activities less than 0.75; *normal*, for activities between 0.75 and 1.25; and *active*, for activities greater than 1.25. Active clays provide the most potential for expansion. Typical values of activities for various clay minerals are as follows:

Mineral	Activity
Kaolinite	0.33 to 0.46
Illite	0.9
Montmorillonite (Ca)	1.5
Montmorillonite (Na)	7.2

3.1.2 Mineralogical Methods

Clay mineralogy is a fundamental factor controlling expansive soil behavior. Clay minerals can be identified using a variety of techniques, the more common of which are listed in Table 3.1.

X-Ray diffraction, the most popular method, works on the principle that beams of X-rays diffracted from crystals are similar to light reflections from the crystal lattice planes. X-Ray analysis is well suited for identification of clay minerals because the wavelength of X-rays is of the same order of magnitude (about 1 Å or 10^{-9} mm) as the atomic plane spacings of these minute crystals. The basal plane spacing is characteristic for each clay mineral group and gives the most intense reflections. Characteristic basal spacings are tabulated in Table 2.3.

TABLE 3.1. Laboratory tests used in identification of expansive soils

Test	Reference	Properties Investigated	Parameters Determined
Atterberg limits	ASTM Standards 1991	Plasticity, consistency	
Liquid limit (LL)	ASTM D-4308	Upper limit water content of plasticity	$PI = LL - PL$ = plasticity index
Plastic limit (PL)	ASTM D-4318	Lower limit water content of plasticity	$LI = \dfrac{w - LL}{LL - PL}$ = liquidity index
Shrinkage limit (SL)	ASTM D-427	Lower limit water content of soil shrinkage	R = shrinkage ratio L_s = linear shrinkage
Clay content	ASTM D-422	Distribution of fine-grained particle sizes	Percent finer than 2 μm
Mineralogical tests	Whittig (1964)	Mineralogy of clay particles	
X-ray diffraction	ASTM STP 479 (1970)	Characteristic crystal dimensions	Basal spacings
Differential thermal analysis	Barshad (1965)	Characteristic reactions to heat treatments	Area and amplitude of reaction peaks on thermograms
Electron microscopy	McCrone and Delly (1973)	Size and shape of clay particles	Visual record of particles
Cation-exchange capacity	Chapman (1965)	Charge deficiency and surface activity of clay particles	CEC (meq/100 g)

Test	Reference	Description	Index
Free swell test	Holtz and Gibbs (1956)	Swell upon wetting of unconsolidated unconfined sample of air dried soil	Free swell $= (V_{wet} - V_{dry})/V_{dry} \times 100\%$
Potential volume change meter	Lambe (1960b)	One-dimensional swell and pressure of compacted, remolded sample under semi-strain controlled conditions	SI (swell index) (lb/ft^2) PVC (potential volume change)
Expansion index text	Uniform Building Code	One-dimensional swell under 1 psi surcharge of sample compacted to 50% saturation initially	Expansion index (EI)
California bearing ratio test	Yoder and Witczak (1975); Kassiff et al. (1969)	One-dimensional swell under surcharge pressure of compacted, remolded samples on partial wetting	Percent swell CBR (%)
Coefficient of linear extensibility (COLE) test	Brasher et al. (1966)	Linear strain of a natural soil clod when dried from 5 psi (33 kPa) to oven dry suction	COLE and LE (%)

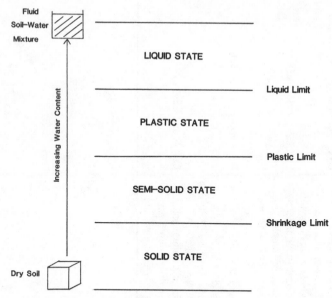

FIGURE 3.1. States of consistency and Atterberg limits of fine-grained soils (Lambe and Whitman, 1969).

Other popular mineralogical methods include differential thermal analysis (DTA) and electron microscopy. DTA consists of simultaneously heating a sample of clay and an inert substance. The resulting thermograms, which are plots of the temperature difference (WT) versus applied heat, are compared to those for pure minerals. Each mineral shows characteristic endothermic and exothermic reactions on the thermograms.

Electron microscopes have provided a means of directly observing the clay particles. Only qualitative identification is possible based on size and shape of the particles using microscopy.

Other mineralogical methods include chemical analysis, infrared spectroscopy, and dye adsorption (Grim, 1968). Radio frequency electrical dispersion has also been used (Basu and Arulanandan, 1973).

3.1.3 Cation Exchange Capacity (CEC)

The CEC is the quantity of exchangeable cations required to balance the negative charge on the surface of the clay particles. CEC is expressed in milliequivalents per 100 grams of dry clay. In the test procedure, excess salts in the soil are first removed and the adsorbed cations are replaced by saturating the soil exchange sites with a known species. The amount of the known cation needed to saturate the exchange sites is determined analytically. The composition of the original cation complex can be determined by chemical analysis of the original extract.

CEC is related to clay mineralogy. High CEC values indicate a high surface activity. In general, swell potential increases as the CEC increases. Typical values of CEC for the three basic clay minerals are as follow (Mitchell, 1976):

Clay Mineral	CEC (meq/100 g)
Kaolinite	3–15
Illite	10–40
Montmorillonite	80–150

The measurement of CEC requires detailed and precise testing procedures that are not commonly done in most soil mechanics laboratories. However, this test is routinely performed in many agricultural soils laboratories and is inexpensive.

3.1.4 Free Swell

The free swell test consists of placing a known volume of dry soil passing the No. 40 sieve into a graduated cylinder filled with water and measuring the swelled volume after it has completely settled. The free swell of the soil is determined as the ratio of the change in volume to the initial volume, expressed as a percentage. A high grade commercial bentonite (sodium montmorillonite) will have a free swell value from 1200 to 2000%. Holtz and Gibbs (1956) stated that soils having free swell values as low as 100% may exhibit considerable expansion in the field when wetted under light loading. Although soils with free swell values below 50% are not considered to exhibit appreciable volume change, Dawson (1953) reported that several Texas clays with free swell values in the range of 50% have caused considerable damage through expansion. It is believed that this was due to extreme climatic conditions in combination with the expansion characteristics of the soil.

3.1.5 Potential Volume Change (PVC)

The soil PVC meter is a standardized apparatus for measuring the swelling pressure of a compacted sample. The PVC meter can be used in the field or laboratory. The test consists of placing a remolded sample into a consolidometer ring with a modified Proctor compactive effort of 55,000 ft-lb/ft^3 (2600 kJ/m^3), at its natural moisture content. The sample is then wetted in the device and allowed to swell against a proving ring. The apparatus is shown in Figure 3.2. The swell index is reported as the pressure on the ring and is correlated to qualitative ranges of potential volume change (PVC) using the chart shown in Figure 3.3 (Lambe, 1960b). The advantages of the test are its simplicity and standardization. However, because the test uses remolded samples the swell index and PVC values are more useful for identification and should not be used as design parameters for in place soils.

3.1.6 Expansion Index Test

The expansion index test was developed in the southern California area in the late 1960s in response to requests from several local agencies for the standardization of testing methods in that area. The method was evaluated statistically by five different testing laboratories in California, and was adopted as a standard by the

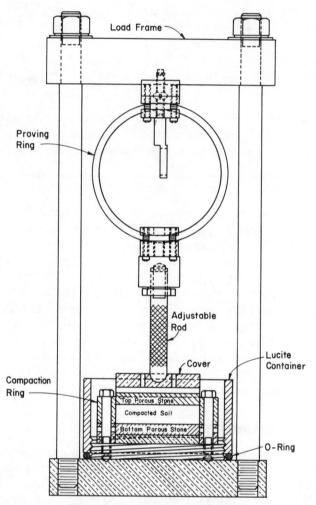

FIGURE 3.2. Potential volume change (PVC) apparatus (Lambe, 1960).

Uniform Building Code (UBC Standard No. 29-2) and many California government agencies. The test is similar to the PVC test, except that swelling is permitted under constant surcharge.

The test consists of breaking down the soil through a No. 4 sieve and bringing the material to approximately optimum moisture content, as determined by ASTM designation D-1557-66T. The soil is "cured" for 6 to 30 hr, and compacted into a standardized, 4-in. (10.2-cm)-diameter mold. The moisture content is then adjusted, if necessary, to bring the sample to approximately 50% saturation. A 144 psf (6.9 kPa) surcharge is applied, and the sample is wetted. Volume change is monitored for up to 24 hr. The expansion index, reported to the nearest whole number, is calculated as follows:

$$EI = 100 \; \Delta h \times F \qquad\qquad (3.2)$$

where Δh = percent swell
$\quad\;\; F$ = fraction passing No. 4 sieve

The expansion potential of the soil is classified according to the expansion index as follows.

EI	Expansion Potential
0–20	Very low
21–50	Low
51–90	Medium
91–130	High
>130	Very high

3.1.7 California Bearing Ratio (CBR)

The CBR test is a penetration resistance test used extensively in highway design methods. The test procedure includes measurement of vertical swell for fine-grained

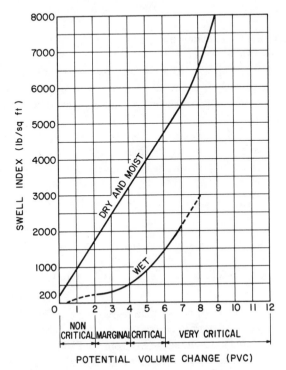

FIGURE 3.3. Swell Index versus PVC (Lambe, 1960).

soils prior to measuring the penetration resistance. Soils are compacted into 6-in. (15.2-cm)-diameter CBR test cylinders at varying moistures and densities, then soaked in water under a surcharge load for 4 days. The surcharge load is chosen to correspond to the static stress to which the soil will be subjected in the field. Swell readings are taken on a dial gage, prior to and after the 4-day soaking period. The samples are allowed to drain for 15 min before the penetration test is performed.

The CBR is a standardized and popular design parameter for highway and airfield pavements. The modified test for expansive soils was developed to assess the affects of swelling on density and strength, and was not intended as an indicator test for identifying potential expansive behavior. However, Kassiff and others (1969) developed charts (Figure 3.4) relating percent swell in the CBR test to PI and percent clay for various moisture–density values. Correlations of this type are specific to the particular soil and project. Kassiff's data were obtained using clays from Israel under a 10 psi (67 kPa) surcharge.

3.1.8 Coefficient of Linear Extensibility (COLE)

The COLE test is a shrinkage test used routinely by the U.S. Soil Conservation Service, National Soil Survey Laboratory for characterizing expansive soils (Brasher et al., 1966). The COLE test determines the linear strain of an undisturbed, unconfined sample on drying from 5 psi (33 kPA) suction to oven dry suction (150,000 psi = 1000 MPa).

The procedure involves coating undisturbed soil samples with a flexible plastic resin. The resin is impermeable to liquid water, but permeable to water vapor. Natural clods of soil are brought to a soil suction of 5 psi (33 kPa) in a pressure vessel. They are weighed in air and water to obtain their volumes using Archimedes' principle. The samples are then oven dried and another volume measurement is performed in the same manner.

COLE is a measure of the change in sample dimension from the moist to dry state and is estimated from the bulk densities of the clod at a suction of 5 psi (33 kPa) and oven dry moisture conditions. The value of COLE is given by

$$\text{COLE} = \Delta L / \Delta L_D = (\gamma_{dD} / \gamma_{dM})^{0.33} - 1 \qquad (3.3)$$

where $\Delta L / \Delta L_D$ = linear strain relative to dry dimensions
γ_{dD} = dry density of oven dry sample
γ_{dM} = dry density of sample at 33 kPa suction

The value of COLE is sometimes expressed as a percent. Whether it is a percent or dimensionless is evident from its magnitude.

COLE has been related to swell index (SI) from the PVC test and other indicative parameters (Franzmeier and Ross, 1968; Anderson et al., 1972; McCormick and Wilding, 1975; Parker et al., 1977; Schafer and Singer, 1976).

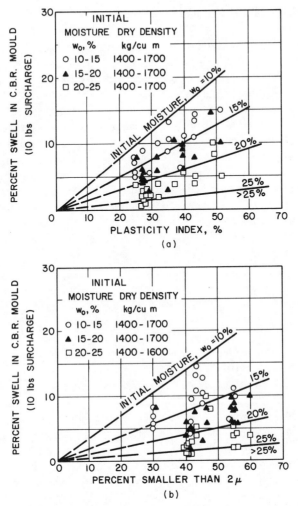

FIGURE 3.4. Percent swell in CBR test versus (a) plasticity index and (b) clay content for Israeli clays (Kassiff et al., 1969).

The National Soil Survey Laboratory (1981) uses Linear Extensibility (LE) as an estimator of clay mineralogy. The ratio of LE to clay content is related to mineralogy as follows:

LE/Percent Clay	Mineralogy
>0.15	Smectites (montmorillonite)
0.05–0.15	Illites
<0.05	Kaolinites

3.2 CLASSIFICATION SCHEMES

The parameters determined from expansive soil identification tests have been combined in a number of different classification schemes. Unfortunately, there has not yet evolved a standard classification procedure, and a different scheme is used in practically every different location.

The most confusing aspect of expansive soil classification is the lack of a standard definition of *swell potential*. Not only do samle conditions vary in the different swell tests used to evaluate swell potential (i.e., remolded or undisturbed), but surcharge loading and other testing factors vary over a wide range of values. For instance, Holtz (1959) defined potential swell as the volume change of an air-dry undisturbed sample when saturated under a 1 psi (6.9 kPa) load. Seed et al. (1962b), however, defined swell potential as the volume change of a remolded sample at optimum moisture and maximum density (Standard AASHTO) under a 1 psi (6.9 kPa) load. Snethen (1979a) provided the following definition of potential swell:

> Potential swell is the equilibrium vertical volume change from an oedometer-type test, expressed as a percent of the original height, of an undisturbed specimen from its natural water content and density to a state of saturation under an applied load equivalent to the in situ overburden pressure.

In all cases, however, the term *swelling potential* refers to the relative capacity for expansion of the different soils.

The amount of swell that may be realized in the field is a function of the environmental conditions. This will include whether the soil is left in place or recompacted. It is important therefore to take into account the fact that two soils may have the same swelling potential, according to their classification, but exhibit very different amounts of swell (Seed et al., 1962b).

Many classification schemes provide an "expansion rating" to provide a qualitative assessment of the degree of probable expansion. Expansion ratings may be something such as "high," "medium," and "low," or "critical" and "noncritical."

The classification schemes provide one or more of the following ratings:

- Ranges of values for either probable percentage of volume change, or probable swelling pressure.
- A qualitative expansion rating, i.e., low, medium, high, and very high expansion potential.
- Some other classification parameter such as CH or A6 (in the USC and AASHTO schemes, respectively). These ratings are specific to the classification scheme, and defined for each individually.

Some of the methods are discussed in the following sections, to serve as examples of the classification procedures.

TABLE 3.2. Expansive soil classification based on colloid content, plasticity index, and shrinkage limit

Data from Index Tests[a]			Probable Expansion (% Total Volume Change)	Degree of Expansion
Colloid Content (% minus 0.0001 mm)	Plasticity Index	Shrinkage Limit		
>28	>35	<11	>30	Very high
20–31	25–41	7–12	20–30	High
13–23	15–28	10–16	10–20	Medium
<15	<18	>15	<10	Low

After Holtz and Gibbs (1956).
[a] Based on Vertical Loading of 1.0 psi.

3.2.1 Soil Classification Methods

Soils are classified in the general schemes; Unified Soil Classification System (USCS) (Howard, 1977) and the American Association of State Highway and Transportation Officials Method (AASHTO) (1978) according to index properties. Soils rated CL or CH by USC, and A6 or A7 by AASHTO, may be considered potentially expansive.

3.2.2 Classification Using Engineering Index Properties

The use of Atterberg limits to predict the swell potential is definitely the most popular approach. Many of the procedures also include clay content. Holtz and Gibbs (1956) presented the criteria shown in Table 3.2 based on undisturbed soil samples. Altmeyer (1955) eliminated the use of percent clay because many laboratories do not include hydrometer analysis in their testing programs. He suggested the use of shrinkage limit or linear shrinkage as shown in Table 3.3. Chen (1965) developed a correlation between percent finer than the No. 200 sieve size, liquid limit, and standard penetration blow counts to predict potential expansion (Table 3.4). He

TABLE 3.3. Expansive soil classification based on shrinkage limit or linear shrinkage

Linear Shrinkage	SL (%)	Probable Swell (%)	Degree of Expansion
<5	>12	<0.5	Noncritical
5–8	10–12	0.5–1.5	Marginal
>8	<10	<1.5	Critical

After Altmeyer (1955).

TABLE 3.4. Expansive soil classification based on percent passing no. 200 sieve, liquid limit, and standard penetration resistance for Rocky Mountain soils

Laboratory and Field Data				
Percentage Passing No. 200 Sieve	Liquid Limit (%)	Standard Penetration Resistance (Blows/ft)	Probable Expansion (% Total Volume Change)	Degree of Expansion
>95	>60	>30	>10	Very high
60–95	40–60	20–30	3–10	High
30–60	30–40	10–20	1–5	Medium
<30	<30	<10	<1	Low

After Chen (1965).

also presented a single index method for identifying expansive soils solely by plasticity index (Table 3.5) (Chen, 1988). Raman (1967) presented the degree of expansion as a function of plasticity index and shrinkage index as shown in Table 3.6.

Seed et al. (1962b), in an extensive study on swelling characteristics of compacted clays, have developed a chart based on activity and percent clay sizes. The chart is shown in Figure 3.5.

Figure 3.6 shows a comparison between the procedures presented by Holtz and Gibbs, Seed et al., and Chen for predicting swell potential as a function of plasticity index (Chen, 1988). These curves show a considerable range of potential volume changes for a given plasticity index. At a plasticity index of 15%, the different methods indicate volume changes of 8.5, 1.5, and 1.0%. Even though the criteria testing was similar for all three procedures, the predicted volume changes cannot be compared directly because the sample conditions and moisture boundaries varied considerably. Holtz and Gibbs' criteria were based on the results from tests on 38 samples that were allowed to swell under conditions from air dry to saturation. The criteria presented by Seed, Woodward and Lundgren were based on testing of remolded soil samples. The results presented by Chen were based on tests on undisturbed samples, which were allowed to swell from natural moisture content

TABLE 3.5. Expansive soil classification based on plasticity index

Swelling Potential	Plasticity Index
Low	0–15
Medium	10–35
High	20–55
Very high	35 and above

After Chen (1988).

**TABLE 3.6. Expansive soil classification
based on plasticity and shrinkage index**

PI (%)	SI (%)	Degree of Expansion
<12	<15	Low
12–23	15–30	Medium
23–32	30–40	High
>32	>40	Very high

After Raman (1967).

to saturation. The differences between the curves can be attributed to different soil types, different initial soil conditions (remolded or undisturbed) and different initial moisture contents (air dried, compacted, or natural). There was no standard by which these swell potential criteria were developed.

Snethen et al. (1977) evaluated seventeen of the published criteria for predicting potential swell. The results of their evaluation showed that liquid limit and plasticity index are the best indicators of potential swell along with natural conditions and environment. A statistical analysis of laboratory data correlating potential swell to 31 independent variables resulted in the classification system shown in Table 3.7. This approach includes consideration of the in situ soil suction, which is an indicator of the natural conditions and environment.

3.2.3   COLE Classification Chart

A new classification scheme, developed by McKeen and Hamberg (1981) and Hamberg (1985), combines engineering index properties with the cation-exchange

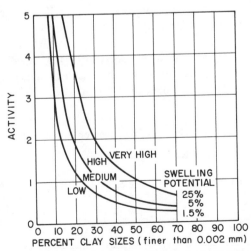

FIGURE 3.5. Classification chart for compacted clays based on activity and percent clay (Seed et al., 1962b, reprinted from Journal of Soil Mechanics ASCE, June 1962 with permission).

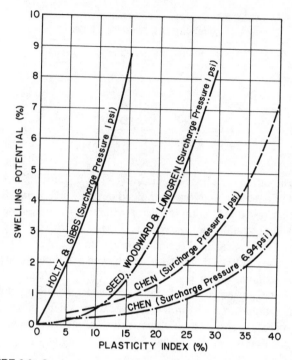

FIGURE 3.6. Comparison of four identification procedures (Chen, 1988).

capacity (CEC). The scheme extended the concepts of Pearring (1963) and Holt (1969), who developed a mineralogical classification chart based on correlations between mineralogy, clay activity (A_c) and a new parameter, cation-exchange activity (CEA_c = CEC/clay content). The Pearring–Holt classification system designated mineralogical groups to certain regions on an A_c versus CEA_c chart, as shown in Figure 3.7.

McKeen and Hamberg (1981) and Hamberg (1985) extended the Pearring–Holt mineralogical classification scheme by assigning COLE values to different regions on the CEA_c versus A_c chart. The method to determine the COLE value is discussed

TABLE 3.7. Expansive soil classification based on liquid limit, plasticity index and in situ suction

LL (%)	PI (%)	μ_{nat},[a]tsf	Potential Swell (%)	Potential Swell Classification
>60	>35	>4	>1.5	High
50–60	25–35	1.5–4	0.5–1.5	Marginal
<50	<25	<1.5	<0.5	Low

After Snethen et al. (1977).

[a] μ_{nat} = soil suction at natural moisture content.

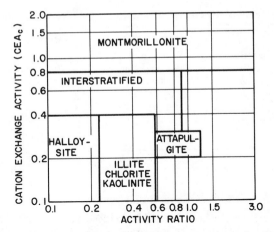

FIGURE 3.7. Mineralogical classification chart based on activity and cation-exchange activity (CED/percent clay) Pearring, 1963; Holt, 1969).

in Section 3.1.8. A new chart was developed based on data obtained from soil survey reports of the U.S. Soil Conservation Service, for soils in California, Arizona, Texas, Wyoming, Minnesota, Wisconsin, Kansas, and Utah. The mineralogical boundaries were adjusted based on the SCS X-ray diffraction results plotted on the chart for CEA_c vs. A_c. The new chart boundaries were drawn for the four mineralogical groups, kaolinite, illite, montmorillonite, and vermiculite. Five mineralogically similar regions were established on the new chart shown in Figure 3.8. The mineralogy of each chart region is summarized in Table 3.8.

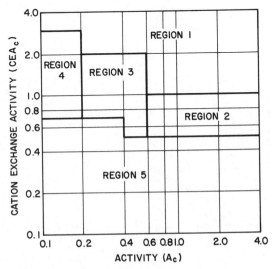

FIGURE 3.8. COLE value classification chart (Hamberg, 1985).

TABLE 3.8. Composition of mineralogical regions on the CEAc-Ac chart

Region	Percentage of Clay Fraction			
	Smectite	Illite	Kaolinite	Vermiculite
1	>50	None	None	None
2	>50	tr[a]–25	tr–25	None
3	5–50	5–25	None	None
4	tr–25	N	10–50	25–50
5	tr	tr–25	10–50	tr

[a]tr, trace < 5%.

The relationship between COLE and clay content was then computed within each of the five chart regions. These relationships for each region are plotted in Figure 3.9. The COLE value can be measured in the laboratory or estimated for a particular soil from the CEA_c versus A_c chart shown in Figure 3.8 and the corresponding relationship for the appropriate region shown in Figure 3.9.

Figure 3.10 was developed to be used as a general classification scheme using the CEA_c versus A_c chart to indicate potentially expansive soils. In general, soils that plot in Regions 1 and 2 will have high to very high expansive potential, moderate potential in Regions 3 and 4, and low expansion potential in Region 5. Figure 3.10 was developed from Figure 3.7 to indicate this classification. Figures 3.9 and 3.10

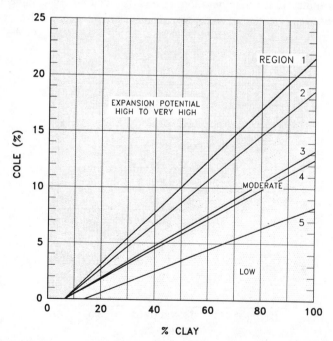

FIGURE 3.9. COLE value as a function of percent clay for regions shown in Figure 3.8.

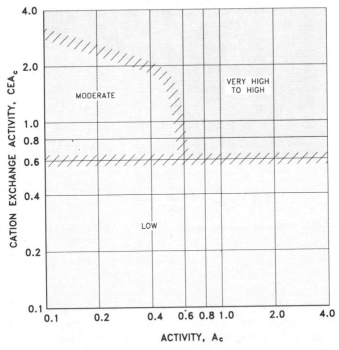

FIGURE 3.10. Expansion potential as indicated by clay activity and CEA.

can be used to determine expansion potential and suction index which will be defined in Chapter 4.

3.2.4 Comparison of Classification Schemes

It is emphasized that the procedures discussed above are only indicators and that the expansion that will occur in the field may vary considerably. Quantitative expansion ratings, including potential values or ranges of values for percent swell and swell pressure, are of little value unless the user is familiar with the soil type and test conditions used to develop the rating criteria. Systems that provide only qualitative ratings, such as high, medium, or low swell potential, may be misused if applied with a design alternative in areas outside the region where the ratings were established. If an empirical, quantitative classification scheme is available for a certain area, it may be applied with some success in design decisions. Otherwise, the classification should be used only to indicate potentially hazardous areas and a need for prediction testing.

It is unfortunate that classification systems are often used as a basis for design selection in practice. This has led to overly conservative construction in some places and inadequate construction in others. Prediction testing and analysis are needed to provide reliable information on which to base design decisions.

4

HEAVE PREDICTION

Several different laboratory procedures for predicting volume changes or swelling pressures of expansive soils exist. Many procedures were developed to reflect specific project and site conditions under investigation at that time. Typically, the procedures reflect the innovativeness of the researcher and frequently include a significant amount of personal opinion.

Common to all soil mechanics analyses for predicting volume changes is the need to define the initial and final in situ stress state conditions. It is also necessary to evaluate the soil moduli or indexes that characterize the stress–strain behavior of each soil type in the profile.

The state of stress must be defined using appropriate stress state variables. Two stress state variables are required to completely define the state of stress in unsaturated soils. The most commonly used variables are one of the two effective stress state variables $(\sigma - u_a)$ or $(\sigma - u_w)$ and the matric suction stress state variable $(u_a - u_w)$ (Fredlund and Morgenstern, 1977). These variables were discussed in more detail in Chapter 2.

Volume changes in expansive soils are generally caused by moisture variations that cause changes in the matric suction stress state variable $(u_a - u_w)$. Methods for measuring matric suction and the volume changes associated with changes in matric suction will be discussed below.

The initial stress state and constitutive properties of a soil can be evaluated using the techniques described in this chapter, but the final stress conditions usually must be assumed. Guidelines are presented that should be of help in defining stress and moisture for selection of final boundary conditions. Nevertheless, each problem will require engineering judgment and consideration of the environmental conditions at each site.

Prediction methods can be separated into three broad categories. These are loosely described as theoretical methods, semiempirical methods, and empirical methods.

Application of any method must rely on testing procedures and analysis techniques developed within a sound theoretical framework. The state of practice continues to include empirical procedures for predicting heave in expansive soils. Those procedures have validity only if they are used within the bounds of soil type, environment, and engineering application for which they were developed. The empirical methods will be discussed for the sake of completeness.

With the development of new concepts in mechanics of unsaturated soil, and practical methods of testing and analysis that incorporate those concepts, the geotechnical engineering community will undoubtedly be making more use of newer techniques. For this reason, theoretically based or semiempirical methods are emphasized over empirical methods.

4.1 CONSTITUTIVE RELATIONSHIPS FOR EXPANSIVE SOILS

A macroscale, phenomenological approach to the study of clay swelling has been applied in most civil engineering applications. Classical soil mechanics uses elastic constitutive (i.e., stress–strain) relationships and principles of continuum mechanics to describe expansive soil behavior. This approach, however, suffered in the past from inadequate definition of the state of stress. A major step in the development of an appropriate framework for macroscale analysis was the definition of the appropriate stress state variables for unsaturated soils.

4.1.1 State of Stress

Most soils that exhibit a significant potential for swelling exist in either an unsaturated condition or, if they are saturated, the pore water pressure is negative. In unsaturated soils, the pore fluid pressure has at least two components, pore water pressure (u_w) and pore air pressure (u_a). In general u_a and u_w are not equal. The pressure difference is balanced at the air–water interface by surface tension forces. This pressure difference is known as capillary pressure or matric suction, μ, and is a valid, independent stress state variable equal to

$$\mu = (u_a - u_w) \tag{4.1}$$

As the value of u_w approaches u_a the suction decreases and the degree of saturation increases. The suction is never allowed to assume a negative value. In cases where the soil is saturated and $u_w > 0$ the value of $(u_a - u_w)$ is always taken as zero.

Early attempts to derive an effective stress equation for unsaturated soils considered the equilibrium of forces at interparticle contacts. The soil suction was viewed as an internal tensile stress acting over a portion of the soil particle surfaces. Empirical parameters were included in equations to include the contribution of $(u_a - u_w)$ to the state of stress. The various equations and parameters that were proposed are summarized in Table 4.1 (Fredlund and Morgenstern, 1977).

TABLE 4.1. Effective stress equations for unsaturated soils

Equation	Description of Variables	Reference
$\sigma' = \sigma - u_a + \chi(u_a - u_w)$	χ = parameter related to degree of saturation u_a = the pressure in gas and vapor phase	Bishop (1959)
$\sigma' = \sigma - \beta' u_w$	β' = holding or bonding factor, which is a measure of the number of bonds under tension effective in contributing to soil strength	Croney et al. (1958)
$\sigma' = \sigma a_m + u_a a_a + u_w a_w + R - A$	a_a = fraction of total area that is air a_m = fraction of total area that is mineral a_w = fraction of total area that is water R, A = repulsive and attractive electrical forces	Lambe (1960a)
$\sigma' = \sigma + \psi p'$	ψ = parameter with values ranging from zero to one p' = pore-water pressure deficiency	Aitchison (1961)
$\sigma' = \sigma + \beta p'$	β = statistical factor of same type as contact area; should be measured experimentally in each case	Jennings (1961)
$\sigma' = \sigma - u_a + \chi_m(h_m + u_a) + \chi_s(h_s + u_a)$	χ_m = effective stress parameter for matric suction h_m = matric suction χ_s = effective stress parameter for solute suction h_s = solute suction	Richards (1966)

After Fredlund and Morgenstern (1977).

To couple both stress state variables into a single effective stress equation is not appropriate. Fredlund and Morgenstern (1977) and Edgar, Nelson, and McWhorter (1989) have shown that the two stress state variables must be considered independently. To couple them in a single equation to define the effective stress would introduce constitutive parameters into the equations of equilibrium and violate the basic laws of mechanics.

Stresses associated with geostatic or mechanical loads may be represented by either one of two variables: $(\sigma-u_a)$ or $(\sigma-u_w)$. Stresses associated with the pore fluid pressures are represented by the matric suction variable (u_a-u_w). The effective stress in terms of either $(\sigma-u_a)$ or $(\sigma-u_w)$ is defined fairly explicitly by the applied loading and knowledge of air and water pressure. It is convenient for most practical applications to use the stress state variable $(\sigma-u_a)$, because the air pressure (u_a) can be taken as zero if continuous air voids exist in the soil. The soil suction (u_a-u_w) can be determined from either the geometry of the problem or from measurements in the field or laboratory.

4.1.2 Constitutive Relationships

Volume changes of an unsaturated soil may be related to the stress state variables using appropriate constitutive relationships. Because the stress state variables are independent, the stress–strain relationships must be depicted on three-axis plots, such as the one shown for void ratio in Figure 4.1. The constitutive surfaces can be linearized by plotting the volume–weight parameters (void ratio, water content or saturation) versus the logarithm of the stress state variables. The constitutive surface in Figure 4.1 may be represented by an equation as follows (Fredlund, 1979):

$$\Delta e = C_t \, \Delta \log(\sigma-u_a) + C_m \, \Delta \log(u_a-u_w) \qquad (4.2)$$

where e = void ratio
      C_t = compression index
      $(\sigma-u_a)$ = saturated effective stress state variable
      C_m = suction index in terms of void ratio and matric suction
      (u_a-u_w) = matric suction

The constitutive relationship for the water phase may be similarly presented:

$$\Delta w = D_t \, \Delta \log(\sigma-u_a) + D_m \, \log(u_a-u_w) \qquad (4.3)$$

where D_t = water content index with respect to saturated effective stress state variable

      D_m = water content index with respect to matric suction

The water content index with respect to matric suction (D_m) is more familiarly known as the moisture characteristic in the discipline of soil physics.

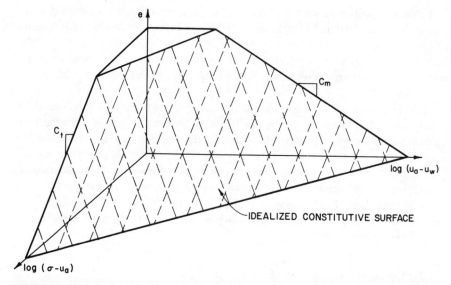

FIGURE 4.1. Idealized three-dimensional constitutive surface for unsaturated soils in terms of void ratio and independent stress state variables.

Measurement of the constitutive parameters requires specialized testing equipment, which allows simultaneous control of air and water pressures and total stress (e.g., Fredlund, 1973; Fargher et al., 1979; Sherry, 1982; Madrid, 1984; Drumright, 1989). Testing methods will be discussed more fully in subsequent sections.

In conventional geotechnical laboratory testing, deformation is measured as a function of stress in the saturated effective stress plane. Techniques for measuring soil suction stresses and volume changes associated with suction changes have been adopted from the field of soil physics.

4.1.3 Suction Indexes

The soil moduli that relate suction changes to volume changes have been defined in several ways, as summarized in Table 4.2. Differences between definitions in Table 4.2 relate to the manner in which strain is represented and to the different components of suction used to define the initial and final stress conditions.

Suction tests at low applied load are simple and do not require specialized equipment. The suction index should be measured within the range of suction changes expected to occur in the field. Most expansive soils exhibit field suctions between about 15 and 1500 psi (100 and 10,000 kPa).

4.2 SOIL SUCTION

The total suction in a soil consists of two parts, osmotic suction and matric suction

$$h = H(h_o, h_c) \tag{4.4}$$

TABLE 4.2. Definitions of suction indexes

Reference	Symbol	Definition	Typical Values	Formation or Soil Type/Location
Fredlund (1979)	C_m	Slope of void ratio versus log matric suction: $C_m = \Delta e / \Delta \log (u_a - u_w)$	$0.1-0.2$ $(C_m \approx C_s)$	Regina clay/Canada
Sherry (1982)	C_m	Slope of void ratio versus log matric suction approximated by $C_m = \Delta e / \Delta \log(u_a - u_w)$	$0.022-0.033$ 0.23	Uranium mill tailings (sand/slimes)/Utah Uranium mill tailings (slimes)/Utah
Johnson (1977, 1979)	C_τ	Slope of void ratio versus log matric suction, approximated by $C_\tau = \alpha G_s / 100 B$ where α = compressibility factor $(0 < \alpha < 1)$ B = slope of suction/water content relationship	0.07 $0.15-0.21$ $0.09-0.23$ $0.07-0.15$ $0.13-0.29$	Loess/Mississippi Yazoo/Mississippi Upper Midway/Texas Pierre Shale/Colo. Marine clay/Sicily
Lytton (1977) McKeen (1981)	γ_h	Slope of volumetric strain versus log total suction: $\gamma_h = \dfrac{(\Delta e / 1 + e_0)}{\Delta \log h}$	$0.02-0.18$ $0.02-0.20$ $0.05-0.22$	Engleford/Texas Yazoo/Mississippi Mancos/New Mexico

(Table continues on p. 64.)

TABLE 4.2. (Continued)

Reference	Symbol	Definition	Typical Values	Formation or Soil Type/Location
Aitchison and Martin (1973) Fargher et al. 1979	I''_{Pt} I''_{Pm} I''_{Ps}	Slope of vertical strain versus log total suction: $$I''_{Pt} = \varepsilon_v/\Delta \log h$$ Slope of vertical strain versus log matric suction: $$I''_{Pm} = \varepsilon_v/\Delta \log(u_a - u_w)$$ Slope of vertical strain versus log solute (osmotic) suction: $$I''_{Ps} = \varepsilon_v/\Delta \log \pi$$	0–0.80 0–0.11 0–0.20	Red clay/Adelaide, S. Australia
Grossman et al. (1968) (U.S.D.A. Soil Conservation Service)	COLE	Value of linear strain corresponding to suction change from 33 kPa (2.53 pF) to oven dry: $$\text{COLE} = \Delta L/L_D = (\gamma_D/\gamma_w)^{1/3} - 1$$ where $\Delta L/L_D$ = linear strain relative to dry dimensions γ_D = bulk density of oven dry sample γ_w = bulk density of sample at 33 kPa suction	0–0.17	Western and Midwestern U.S. soils

where h_o = osmotic suction
  h_c = matric suction

In some publications, the function H has been represented as simply the sum of h_o and h_c. This has not been demonstrated rigorously to be valid. Because of the different phenomena contributing to matric and osmotic suctions it may be questioned whether the two components actually are additive. This is an area requiring more research.

4.2.1 Osmotic Suction

The osmotic suction in a clay results from the forces exerted on water molecules as a result of the chemical activity of the soil. Figure 4.2 illustrates the nature of osmotic suction. In that figure, pure water is shown in contact with a salt solution through a semipermeable membrane, which is a membrane permeable to water molecules but not to the solute. The concentration of the solution causes an attraction to water molecules and hence, a tendency for the flow of water into the solution through the semipermeable membrane. Equilibrium is reached when the hydrostatic pressure head (h_o) of the solution becomes sufficiently large to balance the osmotic forces tending to drive water into the solution. This pressure differential is the osmotic pressure, Ω, as given by

$$\Omega = \rho_s g h_o = RT[C_s] \qquad (4.5)$$

where Ω = osmotic pressure
  ρ_s = solute mass density
  g = gravitational acceleration
  h_o = osmotic pressure head
  R = universal gas constant
  T = absolute temperature
  $[C_s]$ = molar concentration of the solute

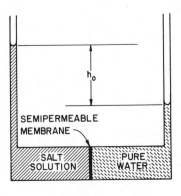

FIGURE 4.2. Development of osmotic pressure across a semipermeable membrane.

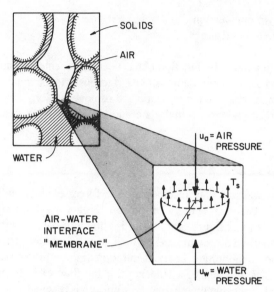

FIGURE 4.3. Air–water interface in soil.

The analogy to this in clay soil behavior is related to the diffuse double layer (DDL) theory discussed in Section 2.2.1.1. Cations are held in DDLs near the clay particle surfaces in concentrations higher than in the bulk solution of the soil pores which is "free" of the adsorptive influences of the particles. This difference in salt concentration generates an osmotic pressure similar to that produced by the semi-permeable membrane shown in Figure 4.2. In addition there may be forces generated by hydration of the salts in very dry soils. These forces result in the suction component called the osmotic pressure head, h_o, or simply osmotic suction.

4.2.2 Matric Suction

Figure 4.3 shows the air–water interface existing in an unsaturated soil. The height above the water table to which the soil will remain saturated will be governed by the pore size and the difference between the air and water pressure. In Figure 4.3, this limiting height is depicted by the bottom of the air channel, which is also shown in an expanded view. Fredlund (1979) discussed the nature of the air–water interface and showed that for engineering purposes the interface can be considered as a *membrane* representing a distinct phase in the soil.

The free body diagram of this membrane is shown in the expanded view. In that figure, r is the radius of an idealized sphere representing the bottom of the air channel, t_s is the surface tension in the membrane, and u_a and u_w are the pressure in the air and water, respectively.

Considering equilibrium of the membrane it can be shown that

$$(u_a - u_w) = \frac{2T_s}{r} \tag{4.6}$$

The term (u_a-u_w) is called the matric suction. In units of pressure head, matric suction is given the symbol h_c, and in units of pressure, μ.

In addition to the surface tension forces there also exist adsorptive forces exerted on the water molecules by the surface of the soil particles. These adsorptive forces account for the fact that the curvature of the water film along the face of the particles is actually in the opposite direction to that of the film between the individual particles as shown in Figure 4.3. The adsorptive forces allow for relatively high tensile stresses to be generated in the soil water than can be significantly greater than one atmosphere.

The bottom of the air channel in Figure 4.3 is shown at the narrowest point between the soil particles. If the difference between the air and water pressure (u_a-u_w) is increased beyond the point where the surface tension can maintain the air–water interface it will "break through" the narrow space between the particles and move down to a point where a smaller space exists. At that point the radius will be smaller and a larger value of suction (u_a-u_w) can be supported as indicated by Eq. (4.6).

If the air pressure is increased or the water pressure is decreased in the soil it will drain until the water has receded into pore spaces that are small enough to support the suction. There is a unique relationship between the water content of the soil and the matric suction. An idealized form of this relationship is shown in Figure 4.4. Obviously, the relationship will depend on the distribution of pores and their sizes, i.e., the soil's fabric and its grain size distribution.

Below the bottom of the air channel in Figure 4.3, the soil is shown as being saturated. As the suction is increased, the magnitude of suction at which the air–water interface "breaks through" and the soil becomes unsaturated is termed the *displacement pressure head*, h_d. The point is labeled in Figure 4.4.

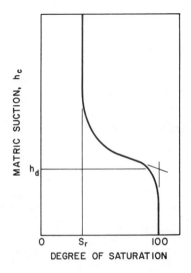

FIGURE 4.4. Water retention curve for soil.

As the matric suction is increased further, the pore water will recede into smaller and smaller pores until a point is reached where an increase in matric suction causes no further decrease in degree of saturation. This value of degree of saturation is termed the *residual saturation*, S_r, also shown on Figure 4.4. To reduce the degree of saturation below S_r requires input of another form of energy, such as heat.

4.2.3 Total Suction

Total suction is a function of both the matric and osmotic suction of the soil. For most practical applications in geotechnical engineering the range of water contents of the soil is such that the adsorbed cations are generally fully hydrated and osmotic forces are fairly constant. Consequently, within the range of water contents encountered in most practical problems, significant changes in osmotic suction do not occur (Krahn and Fredlund, 1972). Changes in total suction that will occur are those that are due only to changes in matric suction. That is

$$\Delta h_o = 0$$
$$\Delta h = H(\Delta h_o, \Delta h_c) = \Delta h_c$$

(4.7)

This is in agreement with the previous discussion that indicated that $(u_a - u_w)$ is a valid stress state variable to define soil behavior. It also suggests that for very dry soils wherein the cation hydration is not complete or the bound water is supersaturated, the total suction would be the appropriate stress state variable to describe soil behavior.

EXAMPLE 3

Given.

Soil profile of clayey fine silt. The water table is at a depth of 20 ft below the ground surface. The displacement pressure head is $h_d = 7.0$ ft. The residual saturation, S_r, of the soil is 28%. Assume continuous pore water for the entire depth below the ground surface.

Find.

Plot matric suction and degree of saturation above the water table.

Solution.

$$u_a = 0$$
$$u_w = -z\gamma_w \quad (z \text{ is the height above the water table})$$
$$(u_a - u_w) = [0 - (-z\gamma_w)] = z\gamma_w$$

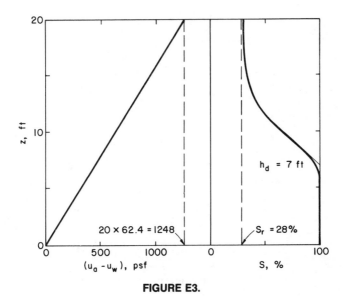

FIGURE E3.

The results are plotted in Figure E3. The exact shape of the curve for degree of saturation cannot be determined from the data given. However, it can be sketched in as being exponential above the displacement pressure head, h_d, and asymptotic to S_r. Below h_d the soil should be saturated.

EXAMPLE 4

Given.

Soil suction given below.

Find.

Depth of the active zone.

Solution.

The suction is plotted against depth in Figure E4. At depths below 20 ft the suction varies linearly and follows a relationship that has a slope corresponding to the unit weight of water. The suction below 20 ft, therefore, is probably in equilibrium with the hydrostatic matric suction above the groundwater table. At depths shallower than 20 ft the suction is much higher than below 20 ft. A discontinuity in the relationship is apparent between 15 and 20 ft. This indicates that the soil above this point has been subjected to forces other than just capillarity, such as desiccation and/or evapotranspiration. Therefore, the depth of the active zone is probably between 15 and 20 ft.

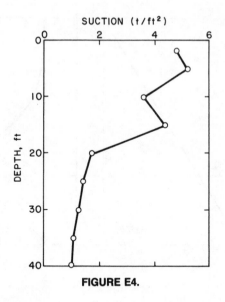

FIGURE E4.

Soil Data for Example 4

Soil	Depth (ft)	Suction (tons/ft²)
Clay, silty, very stiff,	2.0	4.8
moist, brown (CH)	5.0	5.2
	10.0	3.6
	15.0	4.4
	20.0	1.7
	25.0	1.4
	30.0	1.3
	35.0	1.1
	40.0	1.0

4.3 MEASUREMENT OF SOIL SUCTION

The measurement of soil suction can be accomplished by a variety of means. The most commonly used methods include

- tensiometers
- axis translation
- filter paper
- thermocouple psychrometers
- thermal matric potential sensors

Table 4.3 shows some of the features of these methods.

TABLE 4.3. Common laboratory suction measurement methods

Test Method	Suction Component(s)	Range	Advantages	Disadvantages
Tensiometers	Matric	0–12 psi (0–80 kPa)	Commercially available	Low suction range
Axis translation technique (pressure plates)	Matric	0–2000 psi (0–15,000 kPa)	Commercially available	Wetting–drying hysteresis Requires 2 to 10 day equilibration time Requires expensive equipment More appropriate for coarse grained soils
Filter paper sensors	Total	0–150,000 psi[a] (0–10^6 kPa)	Measures full range of suction Inexpensive No specialized equipment required Simple procedure	Requires 7 day equilibration time Care required for moist soils to prevent saturation of sensors
Thermocouple psychrometers (TCPs)	Total	50–150,000 psi[b] (0–10^6 kPa)	Field and laboratory applications Convenient range for most soils Fairly inexpensive equipment, Commercially available	Temperature calibration required Not applicable in very moist soils Requires 2 to 7 days equilibration time
Thermal	Matric	0–30 psi (0–200 kPa)	Commercially available Easy to use	Low suction range

[a]Filter paper accuracy is comparable with TCP accuracy in 50–2000 psi range (McKeen, 1981).
[b]TCP accuracy estimated from reported calibration data (Brown and van Haveren, 1972).

71

In the measurement of suction a means of separating the unsaturated zones from saturated zones is necessary. This is usually accomplished by means of high air entry pressure stones or membranes. Some discussion of these elements is appropriate before discussing methods of measuring suction.

4.3.1   High Air Entry Pressure Stones

The purpose of the high air entry pressure membrane or porous stone is to provide a separation between the air and water phases. The objective is to be able to measure or control water pressure separately from air pressure. This membrane generally consists of a very fine ceramic stone or other material having a very small pore space that will allow the movement of water but will not allow air to permeate at the values of suction being used. The basic principle of operation is to provide for pore spaces that are so small that the displacement pressure is higher than the suction as discussed previously with regard to Figure 4.4.

The displacement pressure required to force air through the stone after it has been saturated is also termed the bubbling pressure, or the air entry pressure. All three of these terms are used synonymously. Stones having a high air entry pressure are frequently referred to as high air entry stones. Ceramic materials having a variety of bubbling pressures can be obtained commercially from several sources. The higher the bubbling pressure, the finer the stone and the lower the permeability. This can have a pronounced influence on the response time of the instrumentation and must be considered in designing a testing program (Fredlund and Morgenstern, 1973).

4.3.2   Tensiometers

Tensiometers consist of a fine porous stone that is placed in contact with the soil. A pressure gauge such as a dial gauge, a manometer, or an electronic transducer is connected to the stone on the side opposite the soil to record the pressure in the water. To maintain the stone in a saturated state so that air passages do not develop, its bubbling pressure must be greater than the soil suction being measured.

The *Quick Draw* tensiometer is a commercially available tensiometer that has been used with reasonable success for measurement of suction in the field. This equipment is available from Soil Moisture Equipment Corporation in Santa Barbara, California. The essential features of this tensiometer are shown in Figure 4.5. The probe, which is about 0.5 m long, is inserted into a precored hole. The vacuum dial gauge is connected to the high air entry porous tip through a small bore capillary tube. An important feature is a null adjusting knob that allows volume changes to be imposed. In this way the time required for equalization of the tensiometer suction with the soil suction can be kept small and disturbance to the in situ suction value is small.

The Quick Draw tensiometer was used to measure suction in the residual soils of Hong Kong for applications in slope stability analyses (Sweeney, 1982). Each reading could be completed in a matter of only a few minutes. Between readings,

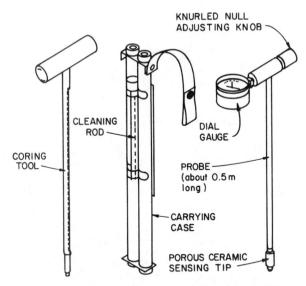

FIGURE 4.5. Quickdraw tensiometer (courtesy of Soil Moisture Equipment Corp.).

the probe was kept saturated in the carrying case. Because the tension is applied to the water in the probe for only a short period of time the potential for cavitation to occur in the tensiometer is minimized. Consequently, it is possible to take numerous readings before deairing of the system is required (Sweeney, 1982).

Flexible tube tensiometers have also been used for more permanent installations both in the field and in the laboratory (Sweeney, 1982; Komurka, 1985).

Because the porous tip allows the migration of salts through the ceramic, tensiometers measure only the matric component of suction. This is adequate for most geotechnical engineering applications because, as pointed out previously, changes in osmotic suction typically are small and relatively large changes in matric suction are the appropriate changes to consider.

Tensiometers are generally limited to measuring suction values significantly less than 1 atmosphere. At suctions greater than this, cavitation of the water in the tensiometer occurs. Figure 4.6 shows data obtained by Sweeney (1982) on Hong Kong soils. It can be seen that the maximum readings obtained were about 90 cbar or 0.9 atmospheres. Very careful procedures are required to read matric suction values of that magnitude with tensiometers. In situ matric suction can be significantly greater than 1 atmosphere in most expansive soils.

4.3.3 Axis Translation

The axis translation technique can be utilized to extend the range of suction that can be measured. The principle on which this technique operates is illustrated using Figure 4.7. This figure shows an idealized sample comprised of soil spheres, water, and continuous air voids. A soil sample is placed over a high air entry ceramic

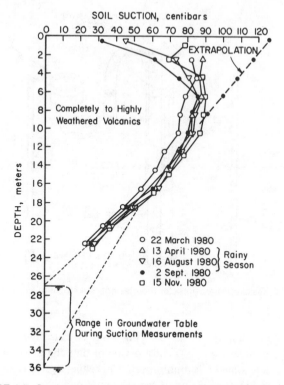

FIGURE 4.6. Suction/depth profiles in Hong Kong soils (Sweeney, 1982).

stone, and air pressure is applied to the sample. Air will not pass through the stone as long as the displacement pressure of the stone is greater than the applied air pressure. The water in the soil will remain connected with the water measuring device through the stone. In this way the water pressure can be maintained throughout the system at any positive value that is desired. Matric suction is the difference between the pore air and pore water pressure. It is not necessary that the water pressure be negative.

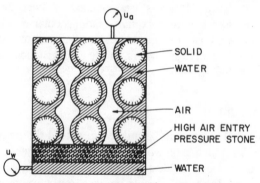

FIGURE 4.7. Axis translation technique.

As the air pressure is increased, the pore water retreats to smaller void spaces in the soil as was discussed in relation to Figure 4.3. Thus, the air and water pressures can be controlled at desired values. Because the matric soil suction is equal to the difference between the air and water pressures, i.e., $(u_a - u_w)$, this provides a means of controlling that stress state variable.

To illustrate this, assume that a soil sample having a matric soil suction of 10 psi (67 kPa) is placed over the high air entry pressure stone in Figure 4.7. If no water was allowed to flow into our out of the sample and the sample were allowed to remain open to the atmosphere (i.e., $u_a = 0$), the water pressure gauge would read -10 psi (-67 kPa). Thus, the suction would be

$$\mu = (u_a - u_w) = 0 - (-10) = 10 \text{ psi} \qquad (4.8)$$

In this case the porous stone is acting as a tensiometer.

If the same sample is placed over the stone in Figure 4.7, and the air pressure is maintained at 30 psi (200 kPa) while the water pressure is increased to 20 psi (133 kPa), the matric suction is again equal to

$$\mu = (u_a - u_w) = 30 - 20 = 10 \text{ psi} \qquad (4.9)$$

The soil suction is the same in these two examples, but the axis of reference against which the water pressure is measured was "translated" from 0 to 30 psi. Thus, this technique is called the axis translation technique.

In the first example [Eq. (4.8)], the water pressure was of such a magnitude that it could be measured by a tensiometer. If, however, the suction had been greater than 1 atmosphere (15 psi or 100 kPa), cavitation in the water pressure measuring line would have made it impossible to measure the negative pore water pressure with a tensiometer. However, by translating the reference axis, both u_a and u_w can be maintained at positive values that can be measured.

This technique is particularly useful for controlling matric suction in laboratory testing programs on unsaturated soils. In triaxial tests or closed cell consolidation tests the bottom stone can be replaced with a high air entry pressure stone and the air pressure can be controlled through the top cap. In this way any stress path in terms of the effective stress state variable $(\sigma - u_a)$ or $(\sigma - u_w)$ and the matric suction $(u_a - u_w)$ can be controlled.

The technique is not as easy to use in the field. However, if good quality undisturbed samples can be obtained they can be transported to the laboratory for suction measurements.

Hilf (1956) proposed the use of this technique to measure the suction of soil samples taken from the field. The apparatus used by Hilf is shown in Figure 4.8. In this apparatus, a saturated high air entry probe was inserted into the soil. The soil suction would tend to draw water into the soil through the porous stone. A null-indicator system was used to sense the tendency for the flow of water into or out of the sample and the air pressure on the sample was increased until no further

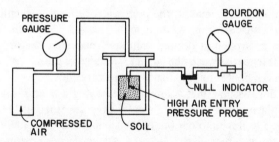

FIGURE 4.8. Hilf's apparatus for measuring soil suction (Hilf, 1956).

tendency for flow existed. When equilibrium had been reached, the value of $(u_a - u_w)$ was equal to the matric suction in the soil.

4.3.4 Pressure Plate Apparatus

Pressure plate tests make use of the axis translation technique. In these tests a chamber is divided by a high air entry pressure stone or membrane as shown in Figure 4.9. Soil samples are placed on top of the high air entry pressure plate and the values of u_a and u_w are controlled. At selected values of soil matric suction, $(u_a - u_w)$, samples are removed and weighed for determination of water content. The results from these tests provide the water retention curve for the soil such as the curve shown in Figure 4.4.

4.3.5 Filter Paper Method

Quantitative ash-free filter papers exhibit a consistent and predictable relationship between water content and suction. Agricultural soil scientists were the first to recognize and use filter papers as indirect soil suction sensors (Gardner, 1937; Fawcett and Collis-George, 1967). Filter paper suction tests have been used routinely by the Water Resources Division of the U.S. Geological Survey (USGS) for many years (McQueen and Miller, 1968). McKeen (1976, 1981), McKeen and Nielsen

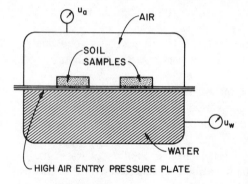

FIGURE 4.9. Pressure plate apparatus.

(1978), McKeen and Hamberg (1981), and Snethen and Johnson (1980) reported the first attempts to use the method for engineering applications.

In this method, a sample of the soil along with a calibrated filter paper is placed in a closed container constructed of noncorrosive material. The filter paper should not be in contact with the soil. The soil sample and filter paper are allowed to equilibrate for a period of at least 7 days at a constant temperature. After the 7-day equilibration period, the filter paper is removed and its water content is determined by precise weighings ($\pm$ 0.0001 g) before and after oven drying.

The filter paper method can be used over a wide range of suction up to approximately 150,000 psi (10^6 kPa). It has been used for a number of investigations of soil water relationships and has been found to produce good results in field investigations (e.g., McKeen, 1980; Snethen and Johnson, 1980; Hamberg, 1985 and others).

The principle on which the filter paper technique is based is that the relative humidity inside the container will be controlled by the soil water content and suction. The filter paper will absorb water until it comes into equilibrium with the relative humidity inside the container. After equilibrium has been reached between the soil water, the filter paper, and the relative humidity in the container, the suction in the filter paper will be at the same value as that in the soil. The humidity, in this case, is influenced by both the osmotic and matric components of the soil suction. Consequently, this technique measures the total suction and not just the matric suction.

Standard quantitative filter papers have a bilinear relationship between suction and water content. Figure 4.10 shows the range of calibration relationships determined

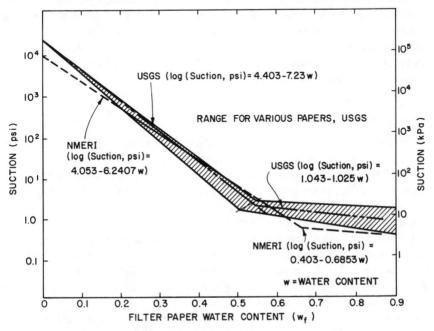

FIGURE 4.10. Filter paper calibration relationships (after McKeen, 1981).

in studies by the USGS (McQueen and Miller, 1968, 1974) and the New Mexico Engineering Research Institute (NMERI) (McKeen and Nielsen, 1978). Calibrations have been determined for different papers, but the most commonly used paper is the Schleicher and Schnell, No. 589, White Ribbon type, or its equivalent.

The calibration technique used in practice should depend on the suction range of interest. The calibration relationship for suction values above about 4.3 psi (30 kPa) represents the moisture retained as a film adsorbed by the fibers of the filter papers (McQueen and Miller, 1974). Calibration in this range is performed by equilibrating the papers above standard salt or acid solutions, for a period of 3 to 7 days. After a sufficient equilibration time, a precise gravimetric moisture content determination is performed for each filter paper. Weights should be measured to the nearest 0.0001 g.

The lower portion of the two-part calibration curve may be found by equilibrating the filter paper sensors with soil samples prepared at known suctions. The pressure plate apparatus is useful for such calibrations.

Most potentially troublesome expansive soils exist at high suction levels. The equilibration between the filter papers and soil samples at high suction levels involves moisture transfer by vapor diffusion processes, and not by capillary action. As noted above, the filter paper suction value represents the total suction. Care must be exercised when testing moist samples to avoid moisture absorption by the filter paper through capillary action from the sample itself or from condensate on the sides of the sample container. McKeen (1981) recommended using two filter paper sensors per soil sample to provide a quality control check.

Comparison of measurements of soil suction using thermocouple psychrometers as discussed in the following section has indicated very good agreement between measurements obtained by these two methods. The particular advantage of the filter paper method is the wide range of suctions over which it can be used and its simplicity. A disadvantage for the use of this method is the degree of accuracy required for weighing the filter paper.

4.3.6   Thermocouple Psychrometers (TCP)

Psychrometers measure relative vapor pressure by meteorological principles. A psychrometer is defined as two similar thermometers with one maintained wet so that cooling from evaporation makes it register a lower temperature than the dry thermometer. The temperature difference is a measure of the dryness of the atmosphere. Spanner (1951) developed a microscale psychrometer for soils work using thermocouples and the principle of the psychrometer.

Thermocouples operate on the principle that an electrical potential will be generated at a junction of two dissimilar metals. The electrical potential will be a function of the temperature. Thermocouples are such junctions. The thermocouple psychrometer (TCP) used in geotechnical engineering is a miniature thermocouple enclosed in a protective casing made of ceramic or steel mesh.

A temperature difference is produced in thermocouple psychrometers (TCPs) using thermocouples in analogy with wet bulb and dry bulb thermometers. A thermocouple sensing junction (wet bulb) may be cooled by passing a small current through it in the proper direction. At the same time, the reference junction (dry bulb) will be heated by the same amount. By constructing the thermocouple reference junctions as massive heat sinks, Spanner (1951) was able to reduce the temperature change at these points so that the reference temperature could be assumed constant. Once the temperature of the sensing junction drops below the dew point of the surrounding atmosphere, water condenses on the sensor, which inhibits further cooling. When the cooling current is stopped, the condensed water evaporates, thus maintaining a lower temperature in the sensing junction compared with the reference junction. A measurable electrical potential is produced by this cooling. The temperature difference associated with the measured potential is correlated with the relative humidity.

Two types of TCPs are available, the major difference between them being the manner of wetting the sensing junction. The Spanner sensor uses Peltier cooling as described above. Another type, developed by Richards and Ogata (1958), requires manual placement of a water drop on the sensing junction. The latter type of sensor is not suitable for field applications.

Operation of TCPs requires a source for a small direct current of 4 to 8 mA and a microvoltmeter in the range of 30 mV to read the TCP output. Commercial psychrometric microvolt-meters are available that have manual cooling switches and variable range voltmeters. It is important to select a uniform procedure for reading output. A standard cooling time period should be selected. A 15-sec cooling period is commonly used. Usually a 2- to 5-sec time lag is recommended between cooling current shut off and reading time (McKeen, 1981).

Calibration of TCPs is essential. Details of calibration and use of TCPs are reported by McKeen (1981), Snethen (1979b), Riggle (1978), and Meyn and White (1972). The devices are calibrated over standard salt solutions. Various concentrations are prepared to obtain humidities in the range of soil suction of interest. Calibration is performed under carefully monitored and controlled temperature conditions, because both electrical potential in the thermocouples and water potential of the salt solutions are highly temperature dependent.

Thermocouple psychrometers have been used successfully for measurement of in situ suction in a number of field investigations (McKeen, 1980; Snethen and Johnson, 1980; Snethen, 1979b; Baker et al., 1973; Peter and Martin, 1973). They can be used for suction values ranging from 15 to 15,000 psi (100 to 100,000 kPa) (Wray, 1984).

4.3.7 Osmotic Method

The concept of osmotic pressure was discussed in Section 4.2.1. Figure 4.11 shows an example of an osmotic tensiometer, developed at the University of Saskatchewan, which makes use of osmotic pressure to measure soil suction. As the water pressure in the high air entry ceramic disk is changed in the presence of soil water, the pressure in the chamber containing the salt solution should change by a corresponding amount. Bocking and Fredlund (1980) presented a detailed discussion of the response

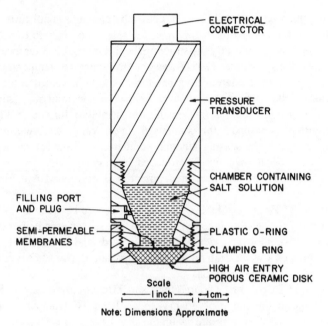

FIGURE 4.11. Osmotic tensiometer.

time, effects of temperature, and gauge compliance on the performance of osmotic tensiometers. They found that the response of osmotic tensiometers was very sensitive to temperature and changes in pore water pressure. Because of this, their use in field applications would be limited. Applications may be possible under more controlled conditions.

4.3.8   Thermal Matric Potential

Thermal matric potential sensors operate on the principle that the thermal conductivity of a porous material is directly proportional to the volumetric water content of that material. By measuring the rate of heat dissipation from a porous ceramic, the water content of the ceramic can be determined. The water content of the ceramic, in turn, is a function of the matric suction. Instrumentation has been developed incorporating a calibrated porous ceramic sensor that is inserted into the soil and allowed to reach equilibrium with the soil. The sensor is calibrated using a pressure plate apparatus (described in Section 4.3.4) to determine the relationship between water content and suction. After equilibrium is reached, a controlled heat pulse is applied in the center of the porous probe, and the increase in temperature is measured over a fixed period of time. The change in temperature is inversely proportional to the volumetric water content of the porous ceramic material. Applying the factory calibration to the change in temperature allows the user to calculate the soil matric suction. The sensor is independent of soil type, dissolved salts, and temperature. A commercially available thermal matric potential sensor is shown in Figure 4.12.

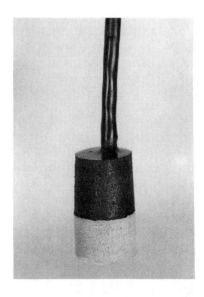

FIGURE 4.12. Thermal matric potential sensor, AGWA II (courtesy of Agwatronics, Inc.).

Lee and Fredlund (1984) compared matric suction values obtained from a thermal matric potential sensor to values obtained using tensiometers and pressure plate apparatus. The results from the sensor agreed well with the results of the other two methods for values up to 30 psi (200 kPa). Above that value, the sensor values showed a considerable amount of scatter. Because expansive soils frequently have matric suction values above 30 psi (200 kPa), care must be taken if using these sensors to ensure that the results are accurate.

4.4 HEAVE PREDICTION BASED ON OEDOMETER TESTS

4.4.1 Consolidation—Swell and Constant Volume Tests

The most common heave prediction tests involve the use of the one-dimensional consolidation apparatus, or oedometer. A wide variety of test procedures have been used, as summarized in Table 4.4. Various loading sequences and applied surcharge pressures have been used in attempts to duplicate in situ conditions. Two basic types of oedometer swell tests may be defined as follows:

The *consolidation-swell test* involves an initial loading of an unsaturated sample to a prescribed stress. The sample is then allowed to swell under that load when water is added. The initial load may represent overburden surcharge, overburden plus structural load, or some other arbitrary surcharge. After swelling, the sample is further loaded and unloaded in the conventional manner. The swell pressure is usually defined as the pressure required to recompress the fully swollen sample back to its initial volume. An idealized plot of consolidation-swell test data is shown in Figure 4.13. In that figure, σ_0' represents the stress at which the sample is wetted and σ_s' represents the swelling pressure according to the above definition. It should

TABLE 4.4. Heave prediction tests using oedometers

Test Name (in chronological order)	Location	Description	Reference
1. Double oedometer method	South Africa	Two tests performed on adjacent samples; a consolidation-swell test under a small surcharge pressure and a consolidation test, performed in the conventional manner but at natural moisture content. Analysis accounts for sample disturbance and allows simulation of various loading conditions and final pore-water pressures	Jennings and Knight (1957)
2. Volumenometer method	South Africa	In specialized apparatus, air-dried samples were inundated slowly under overburden pressure	DeBruijn (1961)
3. Sampson, Schuster, and Budge method	Colorado	Two tests performed on adjacent samples to simulate highway cut conditions; a consolidation-swell test under overburden surcharge, and constant volume-rebound upon load removal test	Sampson et al. (1965)

4. Noble method	Canada	Consolidation-swell tests of remolded and undisturbed samples at various surcharge loads to develop empirical relationships for Canadian prairie clays	Noble (1966)
5. Sullivan and McClelland method	Texas	Constant volume test, samples initially at overburden pressure on inundation	Sullivan and McClelland (1969)
6. Komornik, Wiseman, and Ben-Yacob method	Israel	Constant volume tests at various depths and swell-consolidation tests at various initial surcharge pressures representing overburden plus equilibrium pore water suction, used to develop swell versus depth curves	Komornik et al. (1969)
7. Navy method	United States	Swell versus depth curves determined by consolidation-swell tests at various surcharge pressures representing overburden plus structural loads	Navy (1971)

(*Table continues on p. 84.*)

TABLE 4.4. (Continued)

Test Name (in chronological order)	Location	Description	Reference
8. Wong and Yong method	England	Swell versus depth determined as in (6) and (7), but surcharge loads of overburden plus hydrostatic pore water pressures used	Wong and Yong (1973)
9. USBR method	United States	Double sample test, a consolidation-swell under light load (1 psi), and a constant volume test	Gibbs (1973)
10. Simple oedometer	South Africa	Improved from double oedometer test (1). Single sample loaded to overburden, then unloaded to constant seating load, inundated and allowed to swell, followed by usual consolidation procedure	Jennings et al. (1973)

11. Direct model method (Texas State and Highway Dept)	Texas	Consolidation-swell tests on samples inundated at overburden or end-of-construction surcharge loads	Smith (1973)
12. Mississippi State Highway Dept. method	Mississippi	Consolidation-swell tests on remolded or undisturbed samples, inundated at overburden surcharge loads	Teng et al. (1972; 1973) Teng and Clisby (1975)
13. Controlled strain test	Colorado	Constant volume swell pressure obtained on inundation followed by incremental, strain-controlled pressure reduction	Porter and Nelson (1980)
14. Univ. of Saskatchewan	Saskatchewan, Canada	Constant volume test. Analysis corrects for sample disturbance and apparatus deflection	Fredlund et al. (1980)

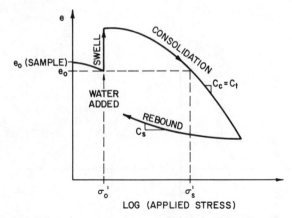

FIGURE 4.13. Typical plot of consolidation-swell test results.

be noted that because the excess pore fluid pressures are allowed to dissipate to zero, it is not necessary to differentiate between effective and total stress in Figure 4.13.

The *constant volume or swell pressure test* procedure involves inundating the sample in the oedometer while preventing the sample from swelling. The swell pressure is reported as the maximum applied stress required to maintain constant volume. Once the swelling pressure stops increasing after soaking, the sample may be rebounded by complete load removal or incremental load removal. Alternatively, it may be loaded beyond the swell pressure and unloaded following the conventional consolidation test procedure. Idealized plots of constant volume test data are shown in Figure 4.14 (Porter and Nelson, 1980).

The analysis of oedometer tests must take into account the loading and wetting sequence, surcharge pressure, sample disturbance, and apparatus compressibility. Only the total stress may be controlled with conventional consolidometers. The matric suction is brought to zero during inundation, but is not measured prior to that. The test results can be interpreted in terms of conventional effective stress theory after the sample has equilibrated with freely available water. However, the initial and final in situ boundary stress conditions must be interpreted in terms of both effective and suction stresses. If oedometer test procedures are used, sample disturbance, and changes in both suction and total stress should be considered in the effective stress analysis.

Two oedometer test procedures have been developed that account for both suction and total stress changes in the stress analysis. In the double oedometer test, developed by South African researchers (Jennings et al., 1973), both initial and final stress conditions are taken into account. A simplified procedure based on the double oedometer test also has been developed.

4.4.2 Double Oedometer and Simplified Oedometer Tests

The simplified oedometer test is a modified consolidation-swell test. The simplified procedure was devised as an alternative to the double oedometer test method, initially proposed by Jennings and Knight (1957). Both procedures are described in this section.

The double oedometer procedure involves testing two, adjacent, undisturbed samples. One sample is consolidated at its natural moisture content. The other sample is inundated while subjected to a small initial load and then consolidated under saturated conditions. Typical results for initially moist and initially dry sample pairs are shown in Figure 4.15.

For two initially moist samples (Figure 4.15a), the curve for the sample tested at natural moisture content is adjusted vertically to match the curve for the saturated sample at high loads. This adjustment compensates for differences in the initial void ratios of the two samples to allow a direct comparison between the samples. The curve for the sample tested at natural moisture content is used to obtain the in situ void ratio, e_0, corresponding to the *total* in situ stress, σ_0. The final void ratio, e_f, is found from the saturated compression curve after calculating the final effective stress.

The change in void ratio during heave is

$$\Delta e = e_f - e_0 \tag{4.10}$$

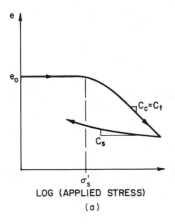

(a)

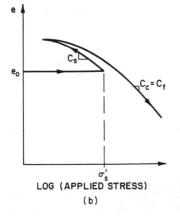

(b)

FIGURE 4.14. Typical constant volume swell test results: (a) sample consolidated beyond swell pressure and (b) sample rebounded from swell pressure, and then consolidated (Porter and Nelson, 1980).

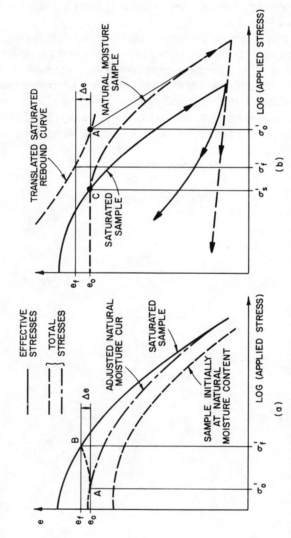

FIGURE 4.15. Double oedometer test results: (a) initially moist sample pair (Jennings and Kerrich, 1962) and (b) initially dry sample pair (Burland, 1965).

where e_0 = initial void ratio corresponding to the initial *total* stress (σ_0) on the natural moisture consolidation curve

e_f = final void ratio corresponding to the final *effective* stress (σ_f) on the saturated consolidation curve

For initially dry sample pairs, the natural moisture content curve is sometimes displaced above the saturated curve at high loads as shown in Figure 4.15b (Burland, 1962). Burland noted that the amount of displacement between the straight line portions of these curves was largely dependent on the initial moisture content. He also observed that the rebound portion of the natural moisture content curve was flat when unloading took place quickly. If the dry sample tests were analyzed in the same manner as the moist sample tests, i.e., by adjusting the natural moisture curves downward to match the virgin compression lines, the heave was overpredicted. Burland (1962) suggested a revised method of analysis. This procedure is shown in Figure 4.15b. The point "A" on this figure is located by the initial in situ effective stress and initial void ratio. The saturated rebound slope is translated upward to pass through "A." The final effective stress is then calculated and the intersection of the translated rebound curve with this value defines the final void ratio.

The simplified procedure evolved as a result of observations made during testing of the natural moisture content samples (Jennings et al., 1973). In the original test, the consolidation of a specimen at natural moisture content was performed solely for the purpose of obtaining the initial condition (e_0, σ_0') (Figure 4.15a). That value provided an estimate of the initial in situ void ratio of the saturated specimen for the prediction of heave. It was recognized that (e_0, σ_0') could be obtained by loading a single specimen to σ_0', at its natural moisture content, then unloading to a light seating load of 0.01 ton/ft^2 (0.14 psi or 1.0 kPa) and performing the saturated swell-consolidation test as usual. The results are illustrated in Figure 4.16a.

Other investigators observed that the slopes of the natural moisture content compression and rebound curves were very flat up to the pressure σ_0' (Burland, 1962; Ralph and Nagar, 1972). Thus, little error was introduced by assuming that the in situ void ratio (e_0) corresponded exactly with the initial sample void ratio e_0 (sample) as shown in Figure 4.16b. This simplified the procedure even further.

Using the simplified oedometer procedure, Jennings et al. (1973) reanalyzed previous results obtained from double oedometer tests (Jennings and Kerrich, 1962). These researchers found that the heave values predicted by the simplified procedure were close to those predicted by double oedometer analysis. Therefore, the simplified test eliminated the uncertainties associated with effects of very dry soils and differences in initial void ratios of sample pairs. The simplified test also eliminated the need to carry tests to very high loads to locate the virgin compression lines.

The simplified oedometer test is tantamount to assuming that the compression curve for the sample with the natural water content is a horizontal line. This is essentially the same as the consolidation-swell test discussed in the previous section and shown in Figure 4.13. For large initial loading values and/or where the curve for the sample at natural water content has significant slope the simplified procedure can underpredict heave. The double oedometer test would be preferred in such cases.

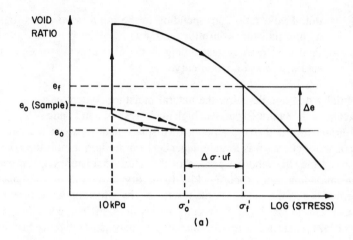

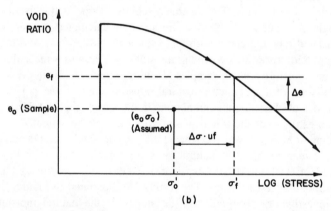

FIGURE 4.16. Simplified oedometer test analysis: (a) single, unsaturated sample loaded to initial in situ stress, then unloaded, saturated and consolidated and (b) analysis assuming e_0 same as initial sample void ratio.

The double oedometer and simplified oedometer test stress paths are shown in Figure 4.17 in terms of the two independent stress state variables. The initial sampling and seating in the oedometer apparatus are shown for both test procedures by stress path segment 0–1–2. Void ratio is assumed to remain constant during sampling and subsequent reloading to the seating stress of 0.01 ton/ft^2 (0.14 psi or 1.0 kPa), as suggested by Jennings et al. (1973).

The simplified oedometer procedure is depicted by stress path 1–2–3–4, where segment 2–3 represents swelling due to soaking under the low seating stress and segment 3–4 represents consolidation at 100% saturation. The consolidation segment 3–4 is shown as a recompression branch of a consolidation curve, since expansive soils are typically highly overconsolidated (Fredlund, 1983).

Idealized stress path 2–0–4 represents the consolidation of a sample at natural moisture content. Void ratio changes are assumed to be very small until the saturated plane is approached. This is the assumption made by Burland (1962) and Ralph and Nagar (1972). The point "C" on Figure 4.17 represents the intersection of the extension of stress path 2–0 and the saturated plane. This point may be considered to represent closely the intersection of the consolidometer curves for the sample at initial water content and the saturated sample (Figure 4.15b). It is also the swelling pressure σ_s' defined in Figure 4.13. The stress at this point, σ_s' is

$$\sigma_s' = \sigma_0' + \Delta\sigma_e' \qquad (4.11a)$$

where $\sigma_0' = \sigma_0 - u_a$
   σ_0 = total overburden stress for initial conditions
   u_a = air pressure

The matric suction equivalent, $\Delta\sigma_e'$, is the change in effective stress $(\sigma - u_a)$, that would be required to restore the sample volume to its original value. In this regard, it is similar to the swelling pressure defined in Figure 4.13. Alternatively, in a constant volume test it would be defined as the change in effective stress, $(\sigma - u_a)$, needed to maintain the sample at its original volume. In this regard, it is similar to the swelling pressure defined in Figure 4.14. In both cases, however, the matric suction equivalent is different from the swelling pressure because it is actually the

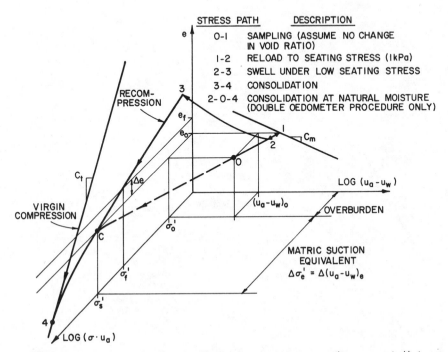

FIGURE 4.17. Idealized double and simplified oedometer test stress paths represented in terms of stress state variables.

change in effective stress from the original state of stress to that at which the matric suction has been brought to zero with no change in volume. Thus,

$$\Delta\sigma_e' = \sigma_s' - \sigma_0' \qquad (4.11b)$$

The void ratio change due to loading and wetting is shown in Figure 4.17 as the difference between the void ratio of the initial sample (e_0) and the final void ratio (e_f) corresponding to the final effective stress plotted on the saturated recompression curve. This was the procedure used in the original double oedometer analysis.

4.4.3 Correction Factors for Oedometer Test Data

The constant volume oedometer test was recommended by Porter and Nelson (1980) and Fredlund (1983) as the best test method to use in predicting expansive soil movements using the concepts presented in Sections 2.2 and 4.1. Using the concept of a matric suction equivalent, as discussed in the previous section, the maximum swell pressure, σ_s', obtained by a conventional constant volume test (Figure 4.14a) theoretically represents the effective stress that would produce the initial volume conditions if there were no soil suction. The effect of sample disturbance on the stress paths can be depicted as shown in Figure 4.18. The points are numbered in

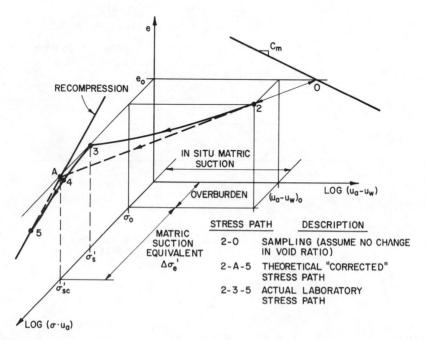

FIGURE 4.18. Idealized and actual laboratory stress paths for the constant volume oedometer test.

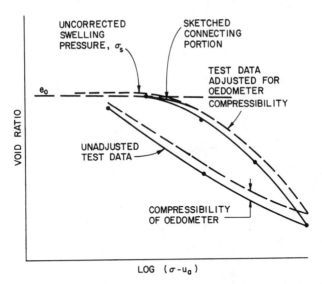

FIGURE 4.19. Correction of constant volume swell test data for apparatus flexibility (Fredlund, 1983).

Figure 4.18 so as to agree with the numbering sequence that will be presented at a later point with reference to Figure 4.20. In Figure 4.18, the initial test conditions are as represented by point 0. During the load to original in situ conditions that sample follows an undefined path between points 0 and 2. At point 2 the sample is inundated and the effective stress increases so as to prevent volume change. If the sample were totally undisturbed it would follow the path 2–A during inundation and A–5 due to subsequent loading. However, due to sample disturbance, the sample follows the path 2–3 during inundation and 3–5 during subsequent loading. Thus, the conventional test underestimates the swelling pressure, σ_s', and it is desired to correct the laboratory curve to determine the corrected swelling pressure, σ_{sc}'.

A method for determining the corrected swelling pressure will be presented. Fredlund (1983) suggests that a correction first should be applied for the compressibility of the consolidation apparatus itself. Apparatus compressibility is significant because desiccated expansive soils are generally highly incompressible and have high pre-consolidation pressures (Fredlund, 1979). The compressibility of the apparatus may be measured with a steel plug substituted for the soil sample. The measured deflections may be subtracted from the test data as demonstrated in Figure 4.19.

To develop the rationale behind the proposed method of correcting the swelling pressure, the void ratio–stress relationship for the sample during its geologic history will first be discussed. In Figure 4.20a the line OMN represents the virgin compression curve for the clay from the time of its deposition. At some time in its geologic history the soil was unloaded along the dashed line M–1. As a result of desiccation or other factors that increased the suction, the void ratio decreased along the line 1–2. It is assumed that during this time the effective stress did not change. The

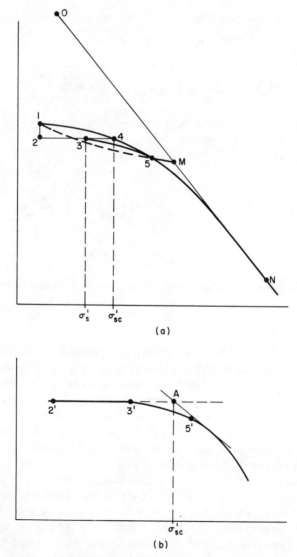

FIGURE 4.20. Correction of constant volume swell test data for sample disturbance: (a) deposition and loading history and (b) determination of corrected swell pressure.

sample was taken from the ground at point 2 and volume changes during sampling are neglected. During a constant volume laboratory test the sample would follow the path 2–3–4. If there was no sample disturbance the measured swelling pressure would be as given at point 4. However, because of sample disturbance, the suction is relaxed more rapidly as previously discussed with regard to Figure 4.18. Consequently, the measured swelling pressure is given by point 3.

As the sample is loaded beyond point 3 the void ratio decreases and the curve approaches the undisturbed recompression line 1–4–5–N. From inspection of Figure

4.20a it is evident that the curvature of the laboratory observed line (2–3–5–N) will be a maximum at point 5 where the disturbed curve rejoins the undisturbed curve. To correct the swelling pressure it is desired to locate point 4.

Figure 4.20b shows only the adjusted part of the curve that is observed in the laboratory tests. The proposed method for correcting the swelling pressure is as follows.

1. Plot the laboratory curve adjusted for oedometer compressibility ($2'-3'-5'$).
2. Locate the point of maximum curvature (point $5'$).
3. Draw a line tangent to the curve along the segment just below point $5'$. Extend this line to where it meets the horizontal extension of line $2'-3'$.
4. The corrected swelling pressure σ'_{sc} is the point of intersection between the tangent to the curve and the horizontal line (point A).

The corrected swell pressure may be significantly greater than the magnitude of the uncorrected swelling pressure.

Once the initial effective stress condition and the corrected swell pressure (σ'_{sc}) have been determined, future deformations may be predicted using the swell index (C_s) from the conventional consolidation-rebound curve, and final effective stress condition. Figure 4.21 represents this procedure for the assumption that the final matric suction value goes to zero. If the rebound surface is unique, the analysis

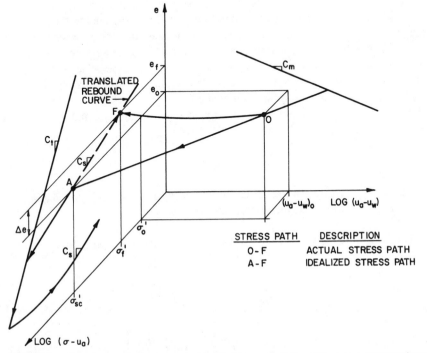

FIGURE 4.21. Actual and idealized stress paths for heave prediction based on constant volume oedometer test.

stress path A–F, starting from (e_0, σ'_{sc}) and following along the saturated effective stress rebound slope (C_s), should arrive at the same final effective stress condition (e_f, σ'_f) as the actual stress path (0–F). The equation for the rebound portion of the oedometer test can be written:

$$e_f = e_0 - C_s \log \frac{\sigma'_f}{\sigma'_{sc}} \tag{4.12}$$

where e_f = final void ratio
 e_0 = initial void ratio
 C_s = swelling index
 σ'_f = final effective stress state
 σ'_{sc} = corrected swelling pressure from constant volume test

It should be noted that because σ'_f is less than σ'_{sc}, log σ'_f/σ'_{sc} will be negative and, hence, e_f will be greater than e_0. The corrected swelling pressure, σ'_{sc}, is determined from the corrected swell test data (Figure 4.20b). It is equal to the sum of the initial effective in situ stress and the matric suction equivalent

$$\sigma'_{sc} = \sigma'_0 + \Delta\sigma'_e \tag{4.13}$$

The final effective stress state must be calculated based on knowledge of the overburden stress, σ'_0, the increment of stress due to applied load, $\Delta\sigma'_{v'}$, and an assumed equivalent effective stress due to the final soil suction. In the oedometer analysis for heave prediction presented above, it is assumed that the final moisture profile approaches 100% saturation. This assumption is necessary because the final stress condition is represented on the saturated stress plane. The matric suction stress state variable is equal to zero on that plane, and the final stress state is appropriately represented by the saturated effective stress state variable $(\sigma - u_w)$. The final effective stress must account for total stress changes and the final pore water pressure

$$\sigma'_f = \sigma'_0 \pm \Delta\sigma - u_{wf} \tag{4.14}$$

where $\Delta\sigma$ = change in total stress due to excavation or surcharge loading
 u_{wf} = estimated final pore water pressure

Total heave is the sum of the displacement in each soil layer. Thus

$$\rho = \sum_{i=1}^{n} \Delta z_i = \sum_{i=1}^{n} \frac{\Delta e_i}{(1 + e_0)_i} z_i \tag{4.15a}$$

or

$$\rho = \sum_{i=1}^{n} \left[\frac{C_s z_i}{(1 + e_0)_i} \log \left(\frac{\sigma'_f}{\sigma'_{sc}} \right)_i \right] \tag{4.15b}$$

where ρ = total heave
$\quad \Sigma z_i$ = heave of layer i
$\quad\quad z_i$ = initial thickness of layer i
$\quad \Sigma e_i$ = change in void ratio of layer i
$\quad\quad\quad$ = $(e_f - e_0)_i$
$\quad\quad\quad$ = $[C_s \log(\sigma_f'/\sigma_{sc}')]_i$
$\quad\quad n$ = number of layers

4.4.4 Prediction of Final Effective Stress Conditions

The final effective stress state must be calculated based on knowledge of the overburden stress, σ_o', the increment of stress due to applied load, $\Delta\sigma_v'$, and the soil suction $(u_a - u_w)$, which may be represented by an equivalent effective stress. In most oedometer analysis procedures, it is assumed that the final moisture profile approaches 100% saturation or at least a condition of zero suction.

Fredlund (1983) suggested that a conservative estimate of heave could be obtained by assuming that the water table would rise to the surface creating hydrostatic positive pore water pressure. Jennings and Kerrich (1962) used the final pore water pressure distribution shown in Figure 4.22a to make heave predictions in South Africa using the double oedometer procedure. This profile was also recommended by other researchers (Johnson and Stroman, 1976; Richards, 1967). It represents the hydrostatic condition for a water table below the surface. For the final pore pressure profile shown in Figure 4.22b the water content distribution would correspond to the appropriate part of the water retention curve, such as shown in Figure 4.4. Nelson and Edgar (1978) showed that during cool seasons thermal gradients under pavements can cause water contents in the subsoils to be significantly higher than those corresponding to the water retention curves. It is believed, therefore, that unless dictated by specific site conditions the most prudent assumption for final suction conditions is to estimate the highest potential water table and then assume that the soil suction can become equal to zero everywhere above that point.

EXAMPLE 5

Given.

The soil profile at a site in Colorado consists of clayshale to a large depth. The water content was observed to vary from about 15% near the surface to about 23% at a depth of 25 ft. Suction data indicated that the active zone is about 20 to 25 ft deep. The swell index, C_s, was measured in oedometer tests to be 0.06. Corrected swelling pressure values were determined by constant volume oedometer tests at different depths. The data are shown below. Assume that after construction of a light slab on the surface, the soil suction goes to zero over the entire height of the active zone.

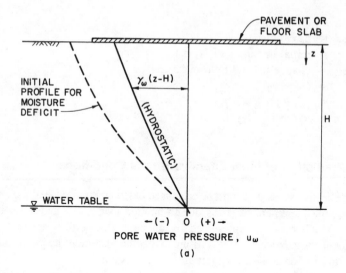

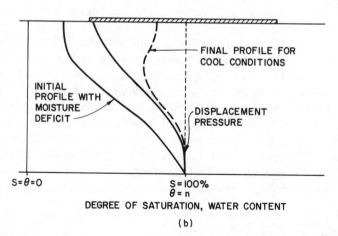

FIGURE 4.22. Moisture profiles above the water table in terms of (a) pore water pressure and (b) water content.

Depth (ft)	Corrected Swell Pressure (psf)	Total Unit Weight (pcf)	Void Ratio	Water Content (%)
4.5	12,250	118	0.65	15.7
13.0	8,500	128	0.62	23.3
18.0	5,500	126	0.64	22.6

Find.

Predict the free field heave that will occur after placement of the slab. Free field heave is the heave due only to change in suction with no change in effective stress.

Solution.

Consider the depth of the active zone to be the entire 25 ft depth. Divide the zone into three layers defined by the data above. The heave of each layer can be computed from Eq. (4.15).

For $(u_a - u_w) = 0$, and for a deep water table,

$$\sigma_f' = z \gamma_{sat}$$

From the above data the value of γ_{sat} for the three layers can be computed to be nearly the same. An average value of 127.2 pcf will be used for the entire height.

Top layer—0.0 to 8.75 ft
At midpoint:

$$\sigma_f' = 4.37 \times 127.2 = 556 \text{ psf}$$

and from Eq. (4.15),

$$\rho_t = \frac{0.06 \times 8.75}{1.65} \log \left(\frac{556}{12,250} \right) = -0.43 \text{ ft}$$

Middle layer—8.75 to 15.5 ft
At midpoint:

$$\sigma_f' = \frac{(8.75 + 15.5)}{2} \times 127.2 = 1542 \text{ psf}$$

and

$$\rho_m = \frac{0.06 \times (15.5 - 8.75)}{1.62} \log \left(\frac{1542}{8,500} \right) = -0.19 \text{ ft}$$

Bottom layer—15.5 to 25.0 ft
At midpoint:

$$\sigma'_f = \frac{(15.5 + 25.0)}{2} \times 127.2 = 2{,}576 \text{ psf}$$

and

$$\rho_b = \frac{0.06 \times (25.0 - 15.5)}{1.64} \log\left(\frac{2576}{5500}\right) = -0.11 \text{ ft}$$

The total heave is the sum of that of the three layers:

$$\rho_{total} = \rho_t + \rho_m + \rho_b = -0.73 \text{ ft} = -9 \text{ in.}$$

Note: Because the argument of the logarithmic term in the above equations is less than 1.0 the computation gives a negative value for ρ. This indicates that the heave occurs in an upward direction. If the applied stress were greater than the corrected swelling pressure the computation would yield a positive value indicating compression of the layer instead of heave.

EXAMPLE 6

Given.

The soil profile and data for Example 5, except that the upper 5 ft of expansive soil will be removed and replaced with compacted nonexpansive soil having a unit weight of 130.0 pcf.

Find.

Predict the free field heave.

Solution.

Consider the active zone to be the same as in Example 5. Consider four layers consisting of a nonexpansive layer 5 ft thick, the lower part of the top layer, and the lower two layers the same as in Example 5.

Topmost layer—0 to 5 ft

$$\rho_{t1} = 0.0 \text{ ft}$$

Top layer—5 to 8.75 ft

$$\sigma'_f = 5.0 \times 130.0 + \frac{(8.75 - 5.0)}{2} \times 127.2 = 888 \text{ psf}$$

and

$$\rho_{t2} = \frac{0.06 \times (8.75 - 5.0)}{1.65} \log\left(\frac{888}{12,250}\right) = -0.16 \text{ ft}$$

For the lower two layers the change in stresses due to the difference in unit weight of the nonexpansive soil is negligible. Therefore, the heave of the middle and bottom layers will be the same as in Example 5. The total heave is

$$\rho_{total} = \rho_{t1} + \rho_{t2} + \rho_m + \rho_b = -0.46 \text{ ft} = -5.5 \text{ in.}$$

EXAMPLE 7

Given.

The soil profile and data for Example 5, except that a large rigid raft foundation 40 ft square having a uniform load of 2000 psf is placed on the surface. Assume that the vertical stress beneath the center of the foundation varies with depth, z, according to the following relationship:

$$\Delta\sigma_v = \frac{2000 \times 40 \times 40}{(40 + z)^2} = \frac{3.2 \times 10^6}{(40 + z)^2} \text{ psf}$$

Find.

Predict the heave at the center of the foundation for the same final water content and suction conditions as Example 5.

Solution.

The solution follows the same lines as for Example 5 except that the final stress condition changed due to the footing load.

Top layer—0.0 to 8.75 ft
At midpoint:

$$\sigma_f' = 556 + \frac{3.2 \times 10^6}{(40 + 4.37)^2} = 2181 \text{ psf}$$

and

$$\rho_t = 0.32 \log\left(\frac{2181}{12,250}\right) = -0.24 \text{ ft}$$

Middle layer—8.75 to 15.5 ft
At midpoint:

$$\sigma_f' = 1542 + \frac{3.2 \times 10^6}{(40 + 12.1)^2} = 2721 \text{ psf}$$

and

$$\rho_m = 0.25 \log\left(\frac{2721}{8500}\right) = -0.08 \text{ ft}$$

Bottom layer—15.5 to 25.0 ft
At midpoint:

$$\sigma_f' = 2576 + \frac{3.2 \times 10^6}{(40 + 20.25)^2} = 6033 \text{ psf}$$

and,

$$\rho_b = 0.35 \log\left(\frac{6033}{5500}\right) = 0.01 \text{ ft}$$

The total heave is

$$\rho_{total} = \rho_t + \rho_m + \rho_b = -0.31 \text{ ft} = -4 \text{ in.}$$

Note: In the bottom layer the applied load causes the stress to be larger than the corrected swelling pressure and that layer has a net compression that subtracts from the heave.

4.4.5 Discussion of Oedometer Tests

The use of oedometer tests to predict heave has the distinct advantage of using conventional testing equipment with which most geotechnical engineers are familiar. It is a rigorous test method if sample disturbance is taken into account. The final stress conditions in these tests assume that the suction has been reduced to zero and as such represents a conservative condition. As long as the practicing engineer recognizes the difference between field and test conditions and has an appreciation for the actual stress state conditions that are represented, these test methods will provide meaningful results.

The use of double oedometer tests to provide parameters for heave prediction has the advantage of simulating field conditions for suction. However, questions regarding the corrections to be applied in order to correlate the two curves in the double oedometer test provide serious limitations. The use of controlled strain (constant volume) tests along with corrections for sample disturbance as described by Fredlund (1983) can provide good quality data. This data used correctly with Eq. (4.15) is believed to provide the most accurate prediction of heave.

4.5 HEAVE PREDICTION BASED ON SOIL SUCTION TESTS

Soil response to suction changes can be predicted in much the same manner as soil response to saturated effective stress changes. The relationship between void ratio and matric suction is analogous to the compression index or swelling index determined by oedometer tests. Heave predictions are made using equations similar to the *reverse consolidation* equations used in oedometer methods. For example, total heave due to changes in both the effective stress and the matric suction may be written as

$$\rho = \sum_{i=1}^{n} \frac{z_i[\Delta e}{(1 + e_0)]_i}$$

(4.16)

$$= \sum_{i=1}^{n} \frac{z_i}{(1 + e_0)_i} [C_{mi} \Delta \log(u_a - u_w) + C_{ti} \Delta \log(\sigma - u_a)]_i$$

where ρ = total heave
    z_i = thickness of layer i
    Δe_i = $(e_f - e_0)_i$
         = $C_{mi} \log[u_a - u_w)_f/(u_a - u_w)_0]_i$
    C_{mi} = matric suction index for layer i
    C_{ti} = effective stress index for layer i
    σ = total stress
    u_a = pore air pressure
    u_w = pore water pressure

The first term in Eq. (4.16) represents the contribution to heave due to changes in suction. If the total stress does not change, only the first term needs to be evaluated. The initial suction conditions may be determined by direct measurement. The final suction conditions must be assumed, as in oedometer analysis procedures. The assumed final suction distribution does not necessarily have to correspond to 100% saturation conditions. Above the water table the soil suction generally will be greater than zero and the soil may or may not be saturated.

4.5.1 U.S. Army Corps of Engineers (WES) Method

An extensive laboratory and field study was conducted by the U.S. Army Corps of Engineers Waterways Experiment Station (WES) to evaluate soil suction and mechanical prediction models for foundation design (Johnson, 1977, 1978, 1980, 1981; Johnson and McAnear, 1973, 1974; Johnson and Snethen, 1978; Johnson et al., 1973). Comparison of laboratory procedures between suction test methods and the oedometer test methods showed that suction tests were simpler, more economical, and more expedient (Johnson, 1977).

The WES prediction method is based on a relationship equivalent to Eq. (4.16). The suction index was not measured directly, but was calculated as follows:

$$C_m = \alpha G_s / 100B \qquad (4.17)$$

where α = compressibility factor (slope of specific volume versus water content curve)

B = slope of suction versus water content curve

G_s = specific gravity of solids

The compressibility factor α relates changes in volume and water content. This parameter may be determined by a test similar to that for shrinkage limit such as the linear shrinkage test (Texas State DHPT, 1970), or the CLOD test described below. Alternatively, α may be estimated from the empirical relationships given by Eq. (4.18) (Croney et al., 1958; Russam, 1965):

$$\begin{aligned}
\alpha &= 0 & PI &< 5 \\
\alpha &= 0.0275\, PI - 0.125 & 5 &< PI \leqslant 40 \qquad (4.18)\\
\alpha &= 1 & PI &\geqslant 40
\end{aligned}$$

In general, the volumetric compressibility factor will decrease with increasing confining pressure. The suction index appears to be related to the swell index, C_s, when α is less than unity and to the compression index, C_c, when α is equal to one. Since highly expansive clays often have compressibility factors equal to or close to unity, substitution of unity for α in the calculation of C_m will be conservative (Johnson, 1977).

Initial suction values and suction versus water content relationships were measured by both thermocouple psychrometers and pressure membrane methods in the WES studies. The following equation was found to adequately represent the suction-water content relationships for numerous clay soils with suctions ranging from 15 to 750 psi (100 to 5000 kPa):

$$\log(u_a - u_w) = A + Bw \qquad (4.19)$$

where $(u_a - u_w)$ = soil suction

w = gravimetric water content

A = intercept

B = slope

The slopes and intercepts (B and A) for a variety of U.S. soils tested at WES are tabulated in Table 4.5.

TABLE 4.5. Constants Relating Suction (kPa) to Gravimetric Water Content (%)

Geologic Formation	State	A	B
Yazoo	Mississippi	7.195	0.107
Hattiesburg	Mississippi	5.721	0.135
Alluvium	Louisiana	5.642	0.088
Prairie Terrace	Louisiana	4.899	0.115
Taylor	Texas	4.658	0.104
Vale	Texas	11.896	0.771
Washita	Oklahoma	8.202	0.394
Hennessey	Oklahoma	10.493	0.527
Chinle	Arizona	5.173	0.188
Chinle	Arizona	7.812	0.245
Mancos	Utah	4.461	0.121
Blue Hill	Kansas	6.575	0.160
Graneros	Kansas	8.381	0.339
Pierre	Colorado	4.953	0.081
Laramie	Colorado	8.434	0.162
Denver	Colorado	7.800	0.314
Mowry	Wyoming	6.403	0.151
Pierre	Wyoming	8.573	0.332
Bearpaw	Montana	8.184	0.339
Pierre	S. Dakota	8.177	0.211

After Snethen et al. (1977).

4.5.2 CLOD Test Method

Volume changes may also be measured using unconfined, undisturbed samples of soil that are coated with a waterproof plastic resin. The original test method utilizing this technique was the coefficient of linear extensibility (COLE) test that is used routinely by the U.S.D.A. Soil Conservation Service (SCS), National Soil Survey Laboratory (Brasher et al., 1966). The COLE test was described in Chapter 3 under "Identification Tests."

The CLOD test is a modification of the COLE procedure. It was developed at the New Mexico Engineering Research Institute for use in heave predictions beneath airfield pavements (McKeen, 1981; McKeen and Hamberg, 1981; McKeen and Nielsen, 1978). Miller and Nelson extended the methodology to predict heave beneath slabs on grade. They applied the method successfully to predict heave beneath actual loaded moisture barriers and building slabs in research conducted at Colorado State University (Hamberg, 1985; Hamberg and Nelson, 1984).

The CLOD test procedure involves coating soil samples with a liquid resin as shown in Figure 4.23. The resin coating allows for volume measurements at different moisture conditions. A suitable resin is DOW Saran F310, which dissolves readily in acetone or methyl ethyl ketone. Once the resin dries on the soil sample, it acts as a flexible membrane, containing the soil material with its natural soil fabric

(a)

(b)

FIGURE 4.23. Sample coating procedure for COLE and CLOD tests. Two or three coats of Saran plastic resin are applied depending on sample porosity.

intact. The resin is essentially waterproof when exposed to liquid water for a short time, but it permits gradual water vapor flow to and from the sample. The volume of a soil sample of any shape may be determined by weighing the soil clod while it is submerged underwater on a balance. The reading of the balance, adjusted for the weight of the pan and water, is a direct measurement of buoyant force on the sample. Sample volume can then be determined by Archimedes' principle.

In the original COLE procedure, each resin-coated sample was brought to 5 psi (33 kPa) suction in a pressure plate device. The sample volumes were determined at the initial adjusted moisture condition using the weighing procedure just described. The samples were then oven dried for 48 hr, followed by another volume determination.

The strain relationships for COLE computation are shown for a cubical soil sample in Figure 4.24. A COLE value was defined as the normal strain that occurs in going from the moist to the dry condition, defined with reference to the dry dimension, as follows (Grossman et al., 1968).

$$\text{COLE} = \frac{L_M - L_D}{L_D} = \frac{L_M}{L} - 1$$

$$= \left[\frac{\gamma_{dD}}{\gamma_{dM}}\right]^{1/3} - 1 \tag{4.20}$$

where L_M = length of moist sample at 33 kPa suction
$\quad L_D$ = length of oven dried sample
$\quad \gamma_{dM}$ = dry density of moist sample at 5 psi (33 kPa) suction
$\quad \gamma_{dD}$ = dry density of oven dried sample

COLE values may be used as a form of suction index because the volume change is always measured over a constant change in soil suction from 5 psi (33 kPa) to oven dry suction.

The CLOD test provides a simpler and more informative test for routine engineering application. The basic difference between the CLOD test and the COLE test procedures is that in the CLOD test, volume changes are monitored along a gradually varying moisture change path. This results in a smooth shrinkage (or swelling) curve for each sample. A moisture characteristic (suction versus water content) curve also may be developed expediently by bringing several samples from a particular stratum to a variety of initial water contents and measuring their suctions prior to the first resin coating.

The basic CLOD test procedure to develop a shrinkage curve is as follows. Coat a sample with resin according to the COLE procedure specifications and measure

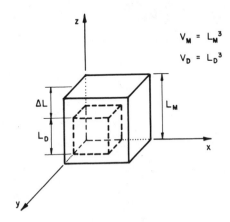

$$V_M = L_M{}^3$$
$$V_D = L_D{}^3$$

ASSUME SAME ΔL IN 3 DIRECTIONS:

$$\text{COLE} = \frac{L_M - L_D}{L_D} = \frac{\Delta L}{L_D}$$

$$\epsilon_V = \frac{\Delta V}{V_D} = 3\frac{\Delta L}{L_D} = 3 \text{ COLE}$$

FIGURE 4.24. Strain relationships for COLE computation.

its volume. Allow sample to dry slowly in air, with periodic volume and weight measurements taken until the sample reaches a constant weight under laboratory humidity conditions. Then oven dry the sample for 48 hr and take a final volume and weight measurement.

These data provide void ratio and water content at various points. Because void ratio and water content are both directly related to soil suction, the relationship between void ratio and water content is tantamount to expressing the effect of suction on void ratio. It is important to note, however, that this is true only at water contents greater than the shrinkage limit. Below the shrinkage limit, changes in water content are not accompanied by changes in volume, by definition. Experiments by Hamberg (1985) showed this relationship to be linear for a silty clay over a range of water contents greater than the shrinkage limit. An idealized shrinkage curve for a clod sample is shown in Figure 4.25 (Hamberg, 1985).

The slope of the curve in Figure 4.25 is designated the CLOD index, C_w, and is analogous to the compressibility factor α described in the previous section. The CLOD index, C_w, is an index of volumetric compressibility with respect to water content.

$$C_w = \frac{\Delta e}{\Delta w} \tag{4.21}$$

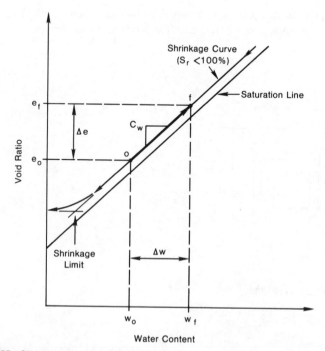

FIGURE 4.25. Shrinkage curve in terms of void ratio and water content for determining the suction modulus ratio, C_w.

Heave, Δz_i for a uniform layer of thickness, z_i, can be determined from Eq. (4.21) as

$$\Delta z_i = \frac{\Delta e}{1 + e_0} z_i = \frac{C_w \Delta w}{1 + e_0} z_i \qquad (4.22)$$

The total heave, ρ, is the sum of all increments of heave for each layer.

$$\rho = \sum_{i=1}^{n} \Delta z_i = \sum_{i=1}^{n} \frac{C_w \Delta w}{1 + e_0} z_i \qquad (4.23)$$

This equation reflects the concept given in Eq. (4.16) but it considers heave only due to suction changes and not changes in effective stress. Initial water content profiles can be measured during the preliminary site investigation. The final water content profile after construction must be predicted on the basis of soil profiles, groundwater conditions, and environmental factors.

In areas of shallow groundwater, it may be reasonable to assume that full saturation with zero suction could develop up to near the surface. In many areas with expansive soil problems, however, the groundwater may be very deep. In arid and semiarid climates, evapotranspiration from the surface causes an initial moisture content profile lower than what can be predicted from hydrostatic equilibrium conditions (Figure 2.2).

When a slab or structure is placed on the surface of a soil, evapotranspiration from the surface is eliminated and water contents increase. Under constant temperature conditions the water content profile should follow the water retention characteristic of the soil as depicted in Figure 4.4. However, because water will flow from warmer areas to cooler areas under thermal gradients, the water content near the surface will increase above the water retention curve during cooler seasons (Edgar et al., 1989). Frost development and frost heave can complicate the situation more but those factors are not considered here.

Experiments conducted at Colorado State University have provided data on seasonal variations of water content profiles under floor slabs (Porter, 1977; Nelson and Edgar, 1978; Goode, 1982; Hamberg, 1985). The soil at the site, located on the Foothills Campus of Colorado State University west of Fort Collins, Colorado, consisted of Pierre Shale to depths greater than 25 ft (8 m).

On the basis of those experiments and measurements of suction on the Pierre Shale using the filter paper method, an idealized initial and final water content profile was developed by Hamberg (1985) as shown in Figure 4.26. Although the active zone was deeper than 6 ft (1.8 m) moisture variation below 6 ft was small at the time of heave prediction.

It was observed that generally the initial water content near the surface did not fall much below the shrinkage limit. Also, the maximum water content under the simulated floor slab did not significantly exceed the plastic limit. Consequently, as shown in Figure 4.26, these values represented the minimum and maximum water contents at the surface for the initial and final water content profiles, respectively. Choosing the shrinkage and plastic limits as initial and final water contents also

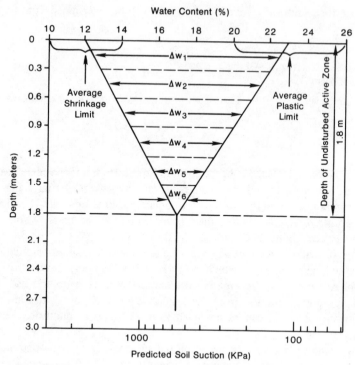

FIGURE 4.26. Idealized moisture boundary profile for the Pierre shale, Fort Collins, Colorado (Hamberg, 1985).

represent lower and upper bounds for heave because virtually all heave takes place between these two water contents. Exceptions to this may occur if osmotic suction is important.

For purposes of heave prediction the water contents were assumed to vary linearly to the zone of constant water content at the bottom of the active zone as discussed with reference to Figures 2.2 and 2.3. It is believed that this represents a sufficiently accurate assumption for the level of confidence within which heave can be predicted.

If desired, the moisture profile could also be depicted in terms of suction as shown by the bottom axis in Figure 4.26. However, the use of water content values rather than suction values to define the boundary conditions causing heave or shrinkage is advantageous for field predictions because water contents are easier to measure than in situ soil suction values. Seasonal variations in water content with depth can be measured to define the upper and lower limits of suction changes and the depth of the "active zone" as discussed in Chapter 2.

The units of suction, shown on the lower axis of Figure 4.26, were converted from the measured water content values using the relationship shown in Figure 4.27. Figure 4.27 was established by filter paper suction measurements, which were discussed in Section 4.3.5. The functional relationship shown in Figure 4.27 represents a statistical correlation between water content and logarithm of suction for values above 15 psi (100 kPa). It was determined on the basis of results of 65 filter paper

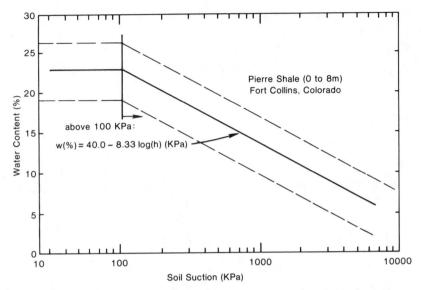

FIGURE 4.27. Relationship between water content and soil suction based on filter paper tests (Hamberg, 1985).

tests (Hamberg, 1985). All field suction values in the Colorado State University investigation were above 15 psi (100 kPa), indicating a strong correlation between water content and suction in the range of field moisture variations. The dashed lines in Figure 4.27 show the range in scatter for the data.

Free field heave (i.e., heave due to suction changes only with no change in effective stress) can be predicted on the basis of water content profiles (Figure 4.26) by dividing the soil profile into incremental layers. Total maximum displacement at the surface can then be predicted by summing the calculated volumetric strains due to the moisture change in each layer across the depth of the active zone using Eq. (4.23).

Example computations for total heave for the idealized moisture profile shown in Figure 4.26 are summarized in Table 4.6. The modulus ratio C_w was assumed to have a constant value of 0.02 based on CLOD test results for the Pierre Shale.

TABLE 4.6. Summary of example prediction computations based on CLOD test results and Figure 4.26[a]

Layer	z (mm)	e_0	C_w	Δw (%)	Δz_i (mm)
1	300	0.9	0.02	10.1	32
2	300	0.8	0.02	8.2	27
3	300	0.7	0.02	6.4	23
4	300	0.6	0.02	4.6	17
5	300	0.55	0.02	2.7	10
6	300	0.5	0.02	0.9	4

[a]$\rho = \Sigma \Delta z_i = 113$ mm.

The total maximum heave predicted at this site was approximately 4 in. (113 mm). This agreed well with observed values of maximum heave for basement floating floor slabs in residential type buildings constructed on Pierre Shale directly adjacent to the test site (Hamberg, 1985).

EXAMPLE 8

Given.

The soil profile and data for Example 2. The CLOD index, C_w, and initial void ratio for the soils are

$$0 \text{ to } 6 \text{ ft: } C_w = 0.012, \ e_0 = 0.65$$

$$6 \text{ to } 10 \text{ ft: } C_w = 0.006, \ e_0 = 0.54$$

$$12 \text{ to } 35 \text{ ft: } C_w = 0.018, \ e_0 = 0.62$$

Find.

Predict the free field heave using the CLOD method.

Solution.

The water content in the active zone can increase to the plastic limit of the soil. Divide the soil into layers between the water content samples, and according to soil boundaries.

0 to 3.7 ft

$$\rho_1 = \frac{0.012 \times (22.0 - 11.0)}{1.65} 3.7 = 0.30 \text{ ft}$$

3.7 to 6.0 ft

$$\rho_2 = \frac{0.012 \times (22.0 - 13.0)}{1.65} 2.3 = 0.15 \text{ ft}$$

6.0 to 10.0 ft

$$\rho_3 = \frac{0.006 \times (12.0 - 5.0)}{1.54} 4.0 = 0.11 \text{ ft}$$

10.0 to 12.0 ft

$$\rho_4 = 0.0$$

12.0 to 15.0 ft

$$\rho_5 = \frac{0.018 \times (26.0 - 16.0)}{1.62} \, 3.0 = 0.33 \text{ ft}$$

15.0 to 20.0 ft

$$\rho_6 = \frac{0.018 \times (26.0 - 19.0)}{1.62} \, 5.0 = 0.39 \text{ ft}$$

$$\rho_{\text{total}} = \Sigma \, \rho_i = 1.28 \text{ ft} = 15 \text{ in.}$$

4.6 EMPIRICAL PROCEDURES

To reduce time and costs of laboratory testing, many investigators have developed empirical relationships for predicting heave. These are usually based on test data from the particular geographic region in which they were developed.

Van der Merwe (1964) developed an easy formula using potential expansiveness and a reduction factor to account for decreasing heave with depth. The expansive soil layer is divided into *n* layers, and the total heave is estimated by Eq. (4.24):

$$\rho = \sum_{i=1}^{i=n} F_i \cdot \text{PE}_i \tag{4.24}$$

where F_i = reduction factor for layer i
   PE_i = potential expansiveness for layer i

F_i is determined from Figure 4.28b.

The value of the potential expansiveness PE is determined by obtaining a classification of the soil from Figure 4.28a based on plasticity index and clay content. That classification is used to determine PE as follows:

Very high	PE = 1 in. per foot depth
High	PE = ½ in. per foot depth
Medium	PE = ¼ in. per foot depth
Low	PE = 0 in. per foot depth

The reduction factor curve was developed for an area in South Africa using double oedometer test results. This method continues to be used in South Africa. However,

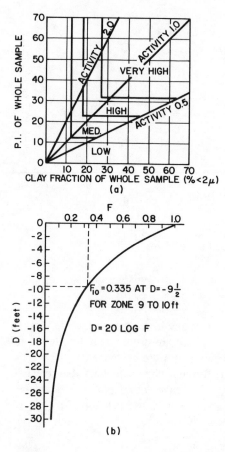

FIGURE 4.28. Relationships for use in Van der Merwe's empirical heave prediction method: (a) potential expansiveness and (b) reduction factor (Van der Merwe, 1964).

it does not consider initial soil conditions such as water content, suction, or density. It should be used only as an indicator of heave, and not for quantitative predictions.

Schneider and Poor (1974) developed statistical relationships for Texas clays between measured swell for various surcharges and the plasticity index and water content. They presented the equations shown in Table 4.7 for predicting the percent swell, S_P.

Many other empirical procedures exist that have not been discussed here. Most are similar to those that have been presented in that they relate the various index or physical properties of the soil to the swell potential.

Many correlations have also been developed that predict the swell pressure that an expansive soil will develop. These procedures were summarized with the classification schemes in Chapter 3.

The empirical methods have the major disadvantage in that they are based on a limited amount of data and actually apply only in the region for which they were developed. Caution should be exercised in their use. Their primary value is as an indicator of expansion potential.

TABLE 4.7. Prediction of percent swell

Surcharge (ft)	Log S_p
0	0.90 (PI/w) − 1.19
3	0.65 (PI/w) − 0.93
5	0.51 (PI/w) − 0.76
10	0.41 (PI/w) − 0.69
20	0.33 (PI/w) − 0.62

After Schneider and Poor (1974).

4.7 DISCUSSION OF HEAVE PREDICTION

Heave prediction can be conducted in ways to imply various degrees of accuracy. Earlier methods predicted heave in terms of "low," "medium," "high," and "very high." Perhaps this should be retained because it does not imply accuracies that are impossible to achieve.

The more quantitative methods of heave prediction that have been presented here can be misleading if too great a degree of accuracy is assumed to exist. It is appropriate to recall the statement quoted in Chapter 2 that "over-refinement of analysis does not lead to improved design" (Dowding, 1979).

It is important to keep in mind the accuracy with which we can predict changes that will occur in the environmental factors contributing to heave. By necessity we must assume the worst case scenarios and attempt to apply the most rigorous theoretical concepts available. Our goal always must be for overprediction of heave instead of underprediction. However, the degree of overprediction must be within reasonable bounds so as to avoid excessive and costly overdesign.

The CLOD method of heave prediction has the distinct advantage of being easy to use and utilizes soil samples having undisturbed fabric and structure. It appears to give good accuracy when compared to actual construction. However, because it does not consider the effective stress it is applicable only for determination of free field heave or heave under very light loads such as pavements or floating floor slabs.

The use of controlled strain tests with corrections for disturbance as presented by Fredlund (1983) (Section 4.4.3) appears to provide the most accurate heave prediction considering effective stress. The primary limitation in that method is the inability to obtain high quality undisturbed samples.

To assess the degree of accuracy inherent with any method is difficult because of the small amounts of data available that compare predicted values with actual heave. Rarely are accurate measurements taken of actual structures where careful prediction analyses are also available. Nevertheless, the authors believe on the basis of discussions with many investigators, that it should be possible to predict actual heave within a few inches if careful sampling and testing are done.

Finally, special care must be exercised to ensure that the testing is done within the stress range that exists and is expected to exist in the field. In the course of interviewing geotechnical engineering consultants in preparation for this book, it was surprising how often the test conditions are specified arbitrarily. Usually, the general stress range does encompass the general range to be expected. In one case, however, the engineer did not know at what seating load their laboratory inundated consolidation-swell test samples! As a profession, the geotechnical engineering community has a responsibility for careful engineering utilizing state–of–the–art technology.

5

DESIGN ALTERNATIVES

5.1 INTRODUCTION

A large number of factors, not all of which are technical, influence foundation types, design methods, and soil treatment practices. Some of these factors include differences in climate, soil conditions, life-styles, mortgage lending practices, and legal standards.

In the northern and central Rocky Mountain areas of North America, for example, the expansive soil problem relates primarily to the upper, desiccated zone of weathered shales and claystones. Because of the cold winter climate, most residences are constructed with basements to accommodate furnaces. Pier and grade beam construction is common in these areas.

In the southern Atlantic and Gulf Coastal Plain states, the active clay zones are somewhat deeper and wetter than those associated with the western shales. Basements are uncommon because of the warmer climate, and stiffened slab-on-grade construction is popular.

Because expansion potential may not manifest itself until months or years after construction, life cycle costs and not just initial costs are very important. This aspect of the risk assessment may affect greatly the selection of a design alternative, because the effort that is practical to apply will be based on the level of funding for the entire project. For this reason alone, funding agencies and/or owners should become involved in the design process at an early stage. Frequently, there is no clear-cut understanding of who is liable for damage that does occur. Consequently, all parties involved must make every effort to keep risks low and come to a good understanding of what the risks are.

Mortgage lending practices and legal standards may have a significant influence on the selection of design alternatives. Often, the foundation design alternative that is optimal for minimizing risk of damage involves costs that are not in line with the anticipated level of funding for the project. As a result, the owners or financiers may prescribe the use of a lower cost foundation system. A lower cost system, in many cases, is a viable alternative but it includes a higher degree of risk to the structure. It is not appropriate for the design engineer to assume those risks on his own or on behalf of the client without making the client aware of the degree of risk involved.

All decision makers, therefore, must be involved in the decision-making process relating to the design. This includes lenders, architects, builders, regulatory officials, engineers, and the eventual owners, the financiers and the principal design firm.

It is particularly important that the owners understand clearly the risks and life cycle costs that are involved with each alternative. Because of all the unquantifiable factors involved in defining risks, the definition of "failure" is often vague. An unacceptable amount of distress may range from purely cosmetic damage to structural failure, depending on personal opinion and the tolerance of the owner. For example, in areas that frequently exhibit distress due to expansive soils, homeowners typically are more tolerant of minor cracking or foundation movement than are homeowners from areas with relatively stable foundation conditions. An agreement about definition of risk, acceptable levels of risk, and costs associated with different levels should be reached by all project participants before construction begins.

An excellent example of the relationship between risk and life cycle costs is illustrated by the author's experience with many of the mining companies in South Africa. Past construction practices on expansive soils for living quarters and offices at mine sites have frequently resulted in excessive maintenance costs. Consequently, these companies have been willing to assume significantly higher initial design and construction costs for more recent construction than one observes for similar projects in North America, for example. These higher initial costs are repaid many times over by the reduction in maintenance costs. In this case the financier, the owner, and the user are the same. Unfortunately, when these parties are all different individuals, the life cycle costs are usually not fully taken into account when design level decisions are made.

5.2 STRUCTURAL FOUNDATION ALTERNATIVES

Foundation design alternatives may be categorized into two types, structural alternatives and soil treatment alternatives. In many cases, soil treatment may be used successfully in conjunction with structural methods.

Two different approaches or design strategies can be used in selecting a foundation design to minimize the differential movement in the superstructure caused by expansive soils. These two strategies are

- Isolate the superstructure from the soil movements, or
- Design a foundation stiff enough to resist differential foundation movement without causing distress to the superstructure.

The major foundation types used in expansive soils are drilled pier and beam systems, reinforced slabs-on-grade, and modified continuous perimeter spread footings. Table 5.1 compares these design alternatives and summarizes the similarities and differences among foundation systems.

The common types of superstructures used with the above foundation systems and their tolerance for deflection are listed in Table 5.2 (Johnson, 1979). Some superstructures such as masonry walls or glass blocks are very intolerant of even relatively small differential movement. A rigid foundation, such as a heavily stiffened mat, or one that will isolate the structure, such as a pier and grade beam foundation, should be used for such structures. Other superstructures, such as a timber construction, can tolerate relatively greater differential movement. A more flexible foundation such as foundations on grade or highly reinforced slabs may be used for more flexible superstructures. Table 5.3 (Johnson, 1979) summarizes the foundation and superstructure systems.

Design methods for the three basic types of foundation systems—pier and grade beam, stiffened slabs-on-grade, and modified continuous perimeter footings—are summarized in Table 5.4. The foundation design alternatives are summarized in terms of design objectives, design procedures, design details, construction quality control, remedial construction measures, and predominant locations where each type of system is used.

5.2.1 Drilled Pier and Beam Foundations

5.2.1.1 *General Description*

Figure 5.1 shows a typical detail of a type of drilled pier and beam foundation system used in the Rocky Mountain Front Range area of Colorado. The grade beam is designed to support the structural load and transfer the distributed stress to the piers. A void space must be maintained beneath the grade beam between the piers. The void isolates the structure from the soil and prevents soil swelling pressures from uplifting the beams. This void space also helps to concentrate the structural load on the piers to counter uplift pressures.

The floor may be constructed as a "floating" slab, which is isolated from the grade beam using a material to minimize friction between the wall and the slab. This system is shown in Figure 5.1. Expansion joint material is often used in the wall/grade beam gap. If a floating floor slab is used, the interior partition walls should be suspended from the upper joists, providing a gap between the bottom of the partition and the top of the slab to allow room for slab movement. The gap may be hidden using a slip joint baseboard. Alternatively, the floor may be designed as a structural slab, supported on the grade beam and isolated from the soil. A wood floor with a crawl space underneath is also a popular option in residential construction.

The piers themselves are typically uncased, reinforced concrete shafts, designed with or without belled bottoms. The main function of the piers is to transfer the

TABLE 5.1. Structural foundation alternatives

Foundation Type	Design Philosophy	Advantages	Disadvantages	Additional Cost	Risk Assessment
Drilled pier and beam (Figure 5.1)	Isolate structure from expansive soil movements	Can be used in a variety of soil types Provides reliable system for soils with high swell potential	Complexity of design and construction Requires specialty contractors	Moderate to high	Low risk for soils of moderate to low expansion potential Low to moderate risk for soils with high expansion potential
Stiffened slab-on-grade (Figure 5.7)	Provide rigid foundation to protect structure from differential soil movement	No specialized equipment required for construction Provides reliable system for soils with moderate swell potential	Not applicable in basement construction Proper design and construction quality control are essential Configuration of building must be relative simple	Low to moderate	Low risk for soils of moderate to low expansion potential Low to moderate risk for soils of high expansion potential

Monolithic wall and slab (Figure 5.15)	Provide rigid foundation to resist differential soil movement	Simple construction No specialized equipment Reduces rotation of foundation wall compared with conventional continuous perimeter footing	Ineffective in highly expansive soils	Low	Low to moderate risk for soils of low expansive potential High risk in soils with high expansion potential
Modified continuous perimeter footings, walls, and basement construction	Provide rigid foundation to resist differential soil movement	Simple construction No specialized equipment	Ineffective in highly expansive soils	None to low	Low to moderate risk for soils of low expansive potential High risk for soils with high expansion potential

TABLE 5.2. Superstructure systems

Superstructure System	Tolerable Deflection/ Length Ratios	Description
Rigid	1/1000	Precast concrete block, unreinforced brick, masonry or plaster walls, slab-on-grade
Semirigid	1/500 to 1/1000	Reinforced masonry or brick, reinforced with horizontal and vertical tie bars or bands made of steel bars, or reinforced concrete beams; vertical reinforcement located on sides of doors and windows; slab-on-grade isolated from walls
Flexible	1/500	Steel, wood framing, brick veneer with articulated joints; metal, vinyl, or wood panels; gypsum board on metal or wood studs; vertically oriented construction joints; strip windows or metal panels separating rigid wall sections with 15-ft. (7.5-m) spacing or less to allow differential movement; all water pipes and drains into structure with flexible joints; suspended floor or slab-on-grade isolated from walls (heaving and cracking of slab-on-grade probable and accounted for in design)
Split construction		Walls or rectangular sections heave as a unit (modular construction); joints at 25-ft (7.5-m) spacing or less between units and in walls; suspended floor or slab-on-grade isolated from walls (probable cracking of slab-on-grade); all water pipes and drains equipped with flexible joints; construction joints in reinforced and stiffened mat slabs at 150-ft (45-m) spacing or less and cold joints at 65-ft (20-m) spacing or less

After Johnson (1979).

TABLE 5.3. Foundation and superstructure systems

Foundation System	Predicted Differential Movement [in. (mm)]	Description	Superstructure System
Shallow	<1/2 (<13)	Continuous wall, individual spread footings	No limit
Reinforced and stiffened slab		Residences and lightly loaded structures; 4-in. (100-mm) reinforced concrete slab on-grade with stiffening beams; 0.5% reinforcing steel; 8- to 12-in.-(200- to 300-mm-) thick beams, external beams thickened and extra steel stirrups added to tolerate high edge forces, as needed; dimensions adjusted to resist loading	Semirigid; flexible; split construction

		Beam Depth, [in. (mm)]	Beam Spacing, [ft (m)]
Light	1/2–1 (13–25)	16–200 (400–500)	20–15 (6.0–4.5)
Medium	1–2 (25–50)	20–24 (500–600)	15–12 (4.5–3.6)
Heavy	2–4 (51–100)	25–36 (600–1000)	12–10 (3.6–3.0)

Foundation System	Predicted Differential Movement [in. (mm)]	Description	Superstructure System
Thick, reinforced mat	No limit	Large, heavy structures; thickness of more than 1 ft (0.3 m)	No limit
Beam on pier	No limit	Underreamed, reinforced, cast-in-place concrete piers; grade beams span between piers about 12 in. (300 mm) above ground level; suspended floors or on-grade first floor isolated from grade beams and walls	No limit

After Johnson (1979).

123

TABLE 5.4. Design summary for major foundation systems used in expansive soils

Foundation Type	Design Objectives	Design Procedures	Design Details	Quality Control	Remedial Measures	Where Used
Drilled pier and beam	Transfer load to stable underlying strata Counteract swelling, uplift with dead load and/or anchoring in nonswelling strata	Establish pier length based on zone of seasonal moisture fluctuation Design diameter based on min. dead load pressure Provide reinforcing steel to resist tension in pier Design grade beam to withstand floor and wall loads	Extend reinforcing steel over entire pier length and tie into grade beam Provide void space beneath structural slabs and beneath grade beam between piers Can provide sleeves to minimize frictional uplift of upper pier	Avoid "mushroom" at top of pier Assure clean pier bottom Maintain void space under grade beam and structural slabs Maintain proper location of piers in line with grade beam Assure cleanliness and proper length of	Excavate void space if not initially provided Underpin and remove improper piers Mudjack (or otherwise level) floating slabs Post tension or install new grade beam	USA: western states Australia Canada India[a] Russia South Africa[a]

124

		Can provide belled pier or shear rings on lower pier to resist uplift Avoid small diameter piers Provide separation between grade beam and floating floor slab	reinforcing steel Avoid segregation of concrete when placing pier Check separation between grade beam and floor slab			
Stiffened slab-on-grade	Provide sufficiently stiff foundation to minimize structural distortions to acceptable levels	Define initial soil heave pattern (mound shape) Define slab loading conditions Assume beam spacing and configuration Determine positive and negative bending moments Compute section modulus and reinforcing requirements	Provide adequate reinforcing steel to ensure beam and slab interaction Adjust beam spacing to coincide with walls or concentrated loads Beam spacing should be less than <20 ft (6 m) to no more than 400 ft^2 (36 m^2) of open area	Standard reinforced concrete inspection Avoid over or undercompaction of soil beneath slab Assure adequate placement depth for anchors in post tension construction	Underpin around edges Mud jacking Epoxy cracks	USA: CA, TX[a] and southern states Australia[a] Canada South Africa[a]

(Table continues on p. 126.)

TABLE 5.4. (Continued)

Foundation Type	Design Objectives	Design Procedures	Design Details	Quality Control	Remedial Measures	Where Used
		Revise section modulus and reinforce as required Calculate beam depth and proportion steel				
Modified continuous perimeter footings, foundation walls, or basement wall "box" construction	Provide foundation wall or beam capable of spanning differential heave Provide sufficiently high heaving pressure to counteract swell pressures	Empirical Minimal design Generally follows local rule of thumb	Can provide void space under portions of foundation wall (to concentrate loads over small areas) For basements, place slab-on-grade as late as possible to allow "rebound" due to excavation	Standard construction inspection	Post tension foundation wall Add adjacent grade beam ("sister wall")	All areas

[a]Major areas of design efforts.

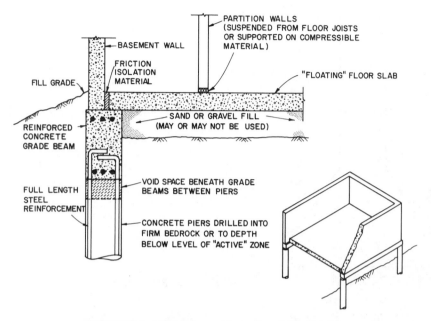

FIGURE 5.1. Drilled pier and beam foundation detail.

structural load to sound bearing or anchorage by a variety of methods. These include (Chen, 1988)

1. Piers drilled into hard bedrock to support high column loads or provide soil anchorage.
2. Friction piers drilled through nonexpansive soil. These may or may not be bottomed on stiff strata or sand.
3. Belled piers bottomed in a stable stratum for supporting medium column loads or providing anchorage.
4. Long piers drilled into a stable zone unaffected by moisture change in swelling soil areas.

Straight shaft or belled bottom piers are used in different areas, depending on the soil conditions, depth to bedrock, and groundwater conditions. In the Front Range area of Colorado, straight shaft piers are common because stable zones of support exist at reasonably shallow depths. The pier diameter is typically kept small, usually between 12 and 18 in. (300 to 450 mm), to minimize the area on which uplift pressures develop. Pier diameters should always be greater than 12 in. (300mm), and preferably larger, to allow for proper placement of concrete along the entire length. Overly small pier diameters result in void spaces, honeycombed concrete, or excessive mixing with soil from the sides of the holes.

Straight shaft piers rely either on minimum dead load pressure on the base or on skin friction in a deeper stable stratum to counteract uplift skin friction in the upper part of the pier.

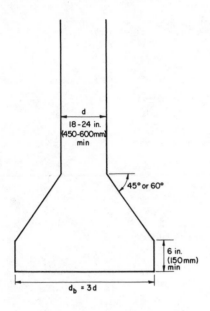

FIGURE 5.2. Belled bottom pier.

Belled bottom, or underreamed, piers are often required in areas where the upper soils are highly expansive, or if there is the possibility of loss of skin friction along the lower anchorage portion of the shaft. Belled piers must have a minimum shaft diameter of 18 to 24 in. (450 to 600 mm) to allow inspection of the bored hole. The ideal bell is shaped like a frustum with a vertical side at the bottom as shown in Figure 5.2. The vertical side should be a minimum of 6 in. (150 mm) high. The sloping sides of the bell are typically formed at either a 60° or 45° angle with the horizontal. Most drillers are capable of forming bells with diameters up to three times the diameter of the shaft (Chen, 1988).

The cost and difficulty of inspection of belled piers are their greatest disadvantages compared with straight shaft piers. However, if there is the possibility of loss of skin friction, particularly due to a rise in the groundwater table, the belled pier design may be advantageous.

5.2.1.2 Design Considerations

The design of drilled piers in compressible soils is based on the applied structural load plus additional downdrag due to settlement of soil adjacent to the pier shaft. These loads are resisted by a combination of end bearing capacity and positive skin friction at lower depths. In contrast, the major concern in expansive soils is the uplift force exerted by soil swelling along the pier shaft within the active zone. Care must be taken in estimating the depth of the active zone when designing drilled piers.

The active zone defined in Chapter 2 was related to climatic factors. In the case of drilled piers, space between the soil and the side of the shaft can provide an access for water to significant depths. This can be particularly critical if water-bearing strata or perched water tables are intersected. If free water gains access to soils below the active zone along the shaft of the pier, deep seated heave can result. Consequently, care should be taken in the design to provide for sealing the space between the soil and the pier.

Two criteria should be considered in the design of drilled piers in expansive soils: upward movement of the top of the pier and the tensile forces developed in the pier. Two different cases will be considered for purposes of establishing design criteria. These are called the rigid pier case, in which deformation of the pier is considered to be zero, and the elastic pier case, in which the pier and the soil are both considered to be elastic.

5.2.1.2.1 Rigid Pier. Chen (1988) and O'Neill (1988) presented similar methods of analysis for rigid piers. The forces acting on a rigid pier are shown in Figure 5.3. Within the active zone, uplift skin friction will be developed. Chen (1988) assumed that this stress is constant throughout the active zone. O'Neill (1988) considered that for a short interval at the bottom of the active zone there will be a transition zone where the uplift skin friction increases from zero (at the bottom) to a limiting constant value that exists throughout the upper part of the active zone. The length of this transition zone, however, is small and not within the accuracy with which the depth of the active zone can be determined. It is prudent, therefore, to assume that the uplift friction is constant throughout the active depth.

The frictional stress f_u, acting between the soil and the pile was defined by Chen (1988) (in the notation used here) as

$$f_u = \alpha_1 \sigma_s' \tag{5.1}$$

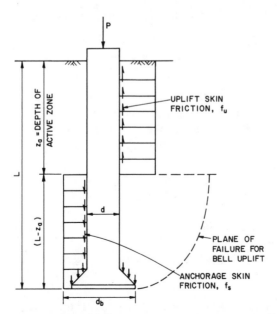

FIGURE 5.3. Forces acting on a rigid pier in expansive soil.

where α_1 = a coefficient of uplift between the pier and the soil

σ_s' = the swelling pressure in terms of effective stresses

O'Neill (1988) expressed the skin friction as

$$f_u = \alpha_2 \sigma_s' \tan \phi_r' \tag{5.2}$$

where α_2 = a factor greater than 1.0 to account for soil structure and disturbance

ϕ_r' = angle of internal friction of the soil for residual strength in terms of effective stress

Equations (5.1) and (5.2) are the same if $\alpha_1 = \alpha_2 \tan \phi_r'$. Chen (1988) reports a value of α_1 equal to about 0.15 based on results of laboratory experiments. O'Neill reported one value of $\alpha_2 = 1.3$, but noted that few data are available on which to base an estimate of α_2. If it is assumed that ϕ_r' varies from about 5° to 10°, $\tan \phi_r'$ would vary from 0.09 to 0.18. Thus, the value of α_1 and $\alpha_2 \tan \phi_r'$ can reasonably be assumed to be between 0.10 and 0.25. For more clayey soils with slow development of heave (so that residual strength is all that is developed) a value of 0.10 may be realistic. If a greater degree of soil–structure interaction exists, a value of 0.25 may be reasonable.

The value of σ_s' can be determined from oedometer tests as discussed in Chapter 4. This should actually be determined on a horizontally oriented sample, but if that is not possible the vertical swell pressure can be used. Normally, the vertical swell pressure will be greater than the horizontal swell pressure. However, in steeply dipping beds, the opposite may be true.

The total uplift force, U, can then be computed by integration of the skin friction, f_u, over the area of the pier within the active zone depth, z_a, as

$$U = \pi d \alpha_1 \sigma_s' z_a \tag{5.3}$$

where d = pier shaft diameter

The uplift force must be resisted by skin friction in the anchorage zone beneath the active zone, uplift resistance of the bell, and applied load.

Chen (1988) expresses this for straight shaft piers as

$$W = \frac{\pi d^2}{4} q_{dl} + \pi d f_s (L - z_a) \tag{5.4}$$

where W = withholding force

q_{dl} = unit dead load pressure

f_s = skin friction below active zone

L = length of pier

The value of f_s must be determined by means similar for evaluation of f_u but taking into account the different environmental conditions within and below the active zone.

O'Neill (1988) suggests using only that portion of the dead load that is applied before structural continuity is established during construction of the superstructure, i.e., kq_{dl} where k is a factor less than 1.0. In his example he used a value of $k = 0.25$ in his computations. The actual value to be used should be based on considerations of the construction sequence.

Setting Eqs. (5.3) and (5.4) equal and using a value of kq_{dl} instead of q_{dl} in Eq. (5.4) gives, for straight shaft piers,

$$L = z_a + \frac{1}{f_s} \left[\alpha_1 \sigma'_s z_a - \frac{kd}{4} q_{dl} \right] \qquad (5.5a)$$

or

$$L = z_a + \frac{1}{f_s} \left[\alpha_1 \sigma'_s z_a - \frac{kP_{dl}}{\pi d} \right] \qquad (5.5b)$$

In the design of a rigid pier it is necessary to assign values to $\alpha_1, f_s, \sigma'_s, k$, and q_{dl}. A diameter of reasonable size can be assumed and Eq. 5.5 solved for L. An iterative process can be used to arrive at a reasonable balance between pier length and diameter.

If belled piers are used, the uplift resistance of the bell must be added to Eq. (5.4). O'Neill (1988) gives the uplift resistance as

$$Q_u = [cN_u + \gamma(L - z_a)]A_u \qquad (5.6)$$

where c = cohesion of the soil
γ = unit weight of the soil
$A_u = \frac{\pi}{4} (d_b^2 - d^2)$

Values of N_u are given in Table 5.5.

If Eq. (5.6) is included in Eq. (5.4) the equation for design of a belled pier is

$$L = z_a + \frac{4\alpha_1 \sigma'_s z_a - dkq_{dl} - d\beta cN_u}{4f_s + d\beta\gamma} \qquad (5.7)$$

where $\beta = (d_b/d)^2 - 1$

TABLE 5.5. Values of N_u for use in Eq. (5.6)

$\dfrac{(L - z_a)}{d_b}$	N_u
1.7	4
2.5	6
≥ 5.0	9

From O'Neill (1988).

Equation (5.7) can be used for design in the same way as discussed for Eq. (5.6).

The methods discussed above do not provide the ability to predict pile movement. It is implicit in the way that the equations are developed that if sufficient anchorage is provided, the movement will be within tolerable limits. Poulos and Davis (1980) provide solutions that allow prediction of movements of a rigid, incompressible pile embedded in an elastic, expansive medium with no soil-pier slip.

As they point out, however, those solutions are of limited validity because some soil-pier slip is certain to exist. It is recommended that prediction of both pier movement and maximum pier load utilize the method described below for an elastic pier and soil with slip.

5.2.1.2.2. Elastic Pier and Soil.
Poulos and Davis (1980) present solutions for pier movement in an elastic medium considering soil-pier slip. These solutions utilize the maximum free field soil heave, stiffness of the soil, stiffness of the pier, and the geometry of the system.

The general case for which solutions are presented by Poulos and Davis (1980) is shown in Figure 5.4. In Figure 5.4 the pier may or may not be belled.

Solutions will be presented for both $d_b/d = 2$ and $d_b/d = 1$. Although drilled piers often have a value of $d_b/d = 3$, solutions were not presented by Poulos and Davis (1980) for that case. The solution for $d_b/d = 2$ will be conservative because the smaller bell will provide less anchorage. Alternatively, an appropriate form of extrapolation can be used from the two cases corresponding to $d_b/d = 1.0$ and $d_b/d = 2.0$.

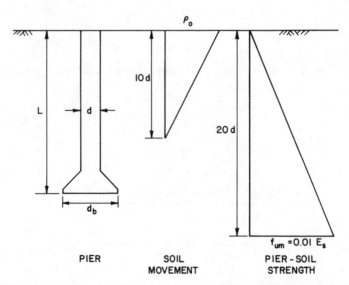

FIGURE 5.4. General case of elastic pier in elastic expansive soil (Poulos and Davis, 1980).

The free field soil heave is assumed to vary linearly from a maximum of ρ_0 at the surface to zero at the depth of the active zone. The free field heave is that which would occur just due to environmental factors with no load applied to the surface. It can be computed using procedures presented in Chapter 4.

The depth of the active zone in Figure 5.4 is assumed to be $10d$. Thus, it would vary from 10 ft for a 1-ft-diameter pier to 20 ft for a 2-ft-diameter pier. This falls within reasonable bounds for most cases. The design charts that will be presented consider a range of depths of active zone.

The pier-soil strength is assumed to equal $0.01E_s$ where E_s is the modulus of elasticity of the soil. Also the base bearing capacity is assumed to vary from $0.36E_s$ at $L = 5d$ to $0.64 E_s$ at $L = 20d$. These values correspond to a cohesionless soil having an angle of internal friction of 30°. As will be shown later, this underpredicts pier load and movement by a factor of about 2 to 3 as compared to a pier-soil strength that is uniform over the length of the pier with a value of $0.005E_s$. If one considers that some cohesion (or adhesion) may exist near the top of the pier and that the angle of internal friction for expansive clays and clayshales is seldom as high as 30°, it is reasonable to expect that the latter case is probably more realistic. Design charts for both cases will be presented. It is recommended that for piers in clays and clayshales the values predicted using the design charts for uniform pier-soil strength be used.

Prior to presenting general design charts the effect of various factors on pier movement and load will be discussed. The sign convention used by Poulos and Davis (1980) is such that negative values of pier movement imply upward movement and negative values of force imply tension.

EFFECT OF PIER LENGTH AND BASE DIAMETER. The load distribution in uniform diameter piers ($d_d/d = 1.0$) is shown in Figure 5.5. For short piers with lengths less than or equal to the depth of the active zone ($z_a = 10d$) the maximum load occurs at a depth of about two-thirds the pier length. For the pier twice as long as the depth of active zone, the maximum load occurs at the bottom of the active zone.

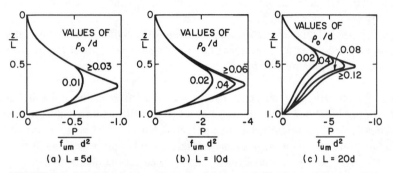

FIGURE 5.5. Load distribution along pier ($d_b/d = 1.0$) (Poulos and Davis, 1980).

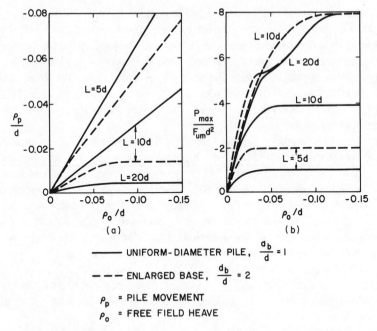

FIGURE 5.6. Effect of pier length and base diameter on (a) pier movements and (b) maximum pier loads (Poulos and Davis, 1980).

Figure 5.6 shows the dimensionless pier movement and load for different pier lengths and for belled and uniform shaft piers. For long piers embedded below the active zone or for belled piers founded just below the active zone a limiting value of pier movement is reached after a certain value of soil movement. This demonstrates the importance of providing adequate anchorage. It also demonstrates that for piers founded well below the active zone, the effect of the bell is minimal.

EFFECT OF PILE SHAFT DIAMETER. Poulos and Davis (1980) show that the effect of pile diameter is unimportant above a certain minimum value. For values of L/d greater than 2.0 the effect of L/d was minimal. Even the movement at $L/d = 10$ was only 20% greater than that at $L/d = 20$.

Thus, a small diameter pier founded well below the active zone can be as effective for reducing heave as a belled pier. The limiting diameter must consider ease of construction and quality control.

EFFECT OF DISTRIBUTION OF SOIL MOVEMENT. The effect of the soil heave profile is shown in Figure 5.7. The curves for cases (i) and (iii) tend to converge at values of $\rho_0/d \leq -0.10$ (absolute value greater than 0.10). Thus, this factor would be of importance primarily for large diameter piers or soils with low to moderate swell potential. In those cases, corrections can be based on the charts shown in Figure 5.7.

EFFECT OF PIER-SOIL STRENGTH DISTRIBUTION. Figure 5.8 shows the effect of pier-soil strength distribution. As discussed previously, the pier-soil strength can be more reasonably expressed by case (i) than case (ii) for high plasticity soils. The design charts presented below will consider both cases. It is recommended that unless soil conditions dictate otherwise, the case for uniform pier-soil strength should be used.

EFFECT OF AXIAL LOAD. The effect of axial load distribution along a pile is often taken into account by adding that load to the forces generated by the soil heave. Superposition of loads is not strictly valid because some soil-pier slip generally occurs. However, Poulos and Davis (1980) showed that the difference between the complete solution did not differ greatly from the solution obtained from superposition.

DESIGN CURVES. Dimensionless curves prepared by Poulos and Davis (1980) are presented in Figures 5.9 and 5.10. These curves can be used to predict maximum pier load and movement for piers in expansive soil. Poulos and Davis (1980) present comparisons between predictions and measurements. The results showed good agreement.

Obviously, the accuracy of the predictions will depend greatly on the quality of the data on which soil parameters are based. With good quality data it should be expected that movements can be predicted within a factor of 2. Loads in the pier should be able to be predicted with even better confidence.

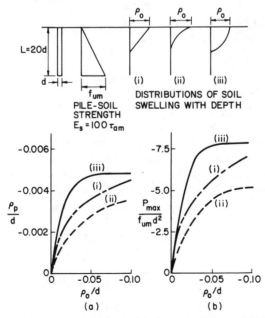

FIGURE 5.7. Influence of soil heave profile on (a) pier movements and (b) maximum pier load (Poulos and Davis, 1980).

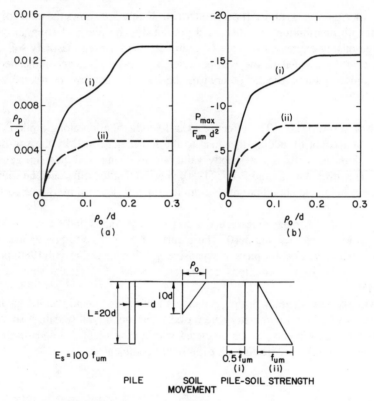

FIGURE 5.8. Effect of pier-soil strength distribution on (a) pier movements and (b) maximum pier load (Poulos and Davis, 1980).

Figure 5.9 considers that the pier-soil strength increases linearly with depth to a maximum f_{um} at the pile tip. Figure 5.10 considers the strength to be uniform with depth equal to f_u along the entire pile length. As discussed above, it is recommended that Figure 5.10 be used unless actual soil conditions dictate otherwise.

In Figures 5.9 and 5.10 the pile load and movement are expressed in the form of a dimensionless parameter, $(\rho_0 E_s/df_{um})$ or $(\rho_0 E_s/df_u)$ for different dimensionless depths of the active zone, z_a/L. All of the variables have been defined previously.

The maximum load, P_{max}, is presented as a ratio of the load, P_{FS}, that would occur if full adhesion was mobilized along the entire shaft. Thus,

$$P_{FS} = \int_0^L (f_u \pi d) dz \qquad (5.8)$$

For the case where the pier-soil strength increases linearly with depth,

$$P_{FS} = -\frac{1}{2} f_{um} \pi d L \qquad (5.9)$$

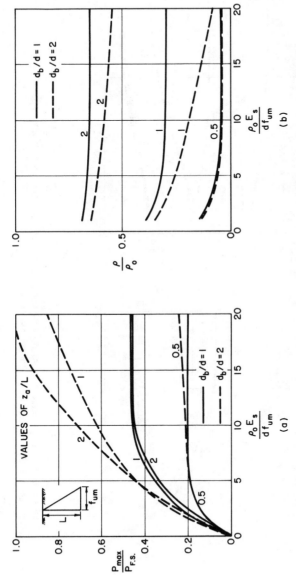

FIGURE 5.9. Design chart for piers in expansive soil—linearly increasing pier-soil shear strength with depth (Poulos and Davis, 1980: Modified by H. Poulos, 1991).

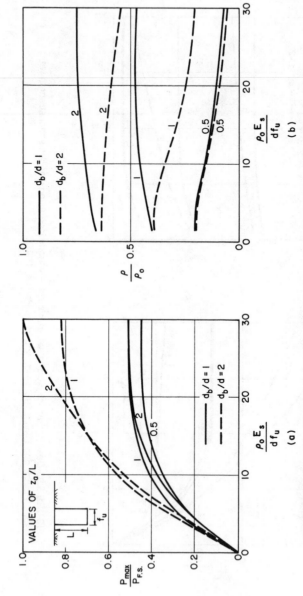

FIGURE 5.10. Design chart for piers in expansive soil—uniform pier-soil strength with depth (Poulos and Davis, 1980: Modified by H. Poulos, 1991).

If the pier-soil strength is uniform over the length of the pier,

$$P_{FS} = -f_u \pi dL \tag{5.10}$$

As indicated in Figure 5.5 the maximum load P_{max} occurs at or near the bottom of the active zone for piers imbedded well below the active zone.

DETERMINATION OF PIER MOVEMENT AND LOAD DUE TO AXIAL LOADING. Poulos and Davis (1980) present solutions for piers in an elastic medium subjected to axial loads. Concrete piers used in expansive soil applications generally fall within a fairly narrow range of sizes and conditions. Therefore, it is possible to replot their data to provide design charts to allow for consideration of axial loads.

These charts have been prepared for concrete piers having solid cross section (i.e., not hollow) and for a ratio of concrete to soil stiffness of 1000 or greater.

Figure 5.11 shows the pier movement in the form of a dimensionless parameter $(\rho_p E_s d)/P_{dl}$. P_{dl} is the axial dead load on the pier, and ρ_p is the pile movement.

Figure 5.12 gives the load at the base of the pier P_b as a ratio of P_{dl}. The load can be assumed to vary linearly with depth. The maximum load due to the expansion will occur at the bottom of the active zone. The load due to the axial load can be determined by interpolation and added to that due to expansion.

Generally, the movement due to expansion will be upward (i.e., negative) while the movement due to axial load will be downward (i.e., positive). Also the load

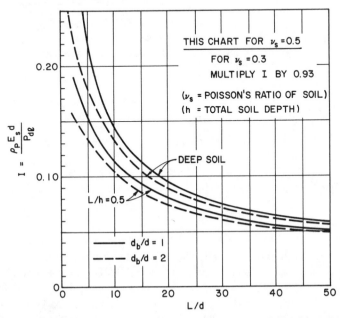

FIGURE 5.11. Design chart for predicting pier movement due to axial load (constructed using data from Poulos and Davis, 1980).

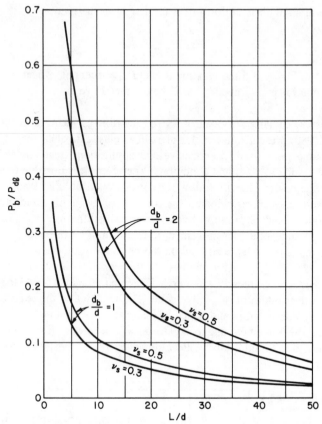

FIGURE 5.12. Design chart for predicting load distribution due to axial load (constructed using data from Poulos and Davis, 1980).

due to expansion will be tension (i.e., negative) while the load due to axial load will be compression (i.e., positive).

EXAMPLE 9

Given.

A pier and grade beam foundation will be utilized in a deep deposit of clayshale. The depth of the active zone is 20 ft. The corrected swelling pressure of the clayshale is 7400 psf, and the swelling modulus, $C_s = 0.038$. The clayshale has a saturated unit weight of 130 pcf and a void ratio of 0.62 throughout the entire depth. The coefficient of uplift, $\alpha_1 = 0.16$. Below the active zone the skin friction between the soil and the pier is 800 psf. The minimum dead load on the pier is 30,000 lb, all of which can act on the pier ($k = 1.0$). Assume that the soil suction can go to zero over the entire depth of the active zone after construction, but that no water table develops.

Find.

a. Required length and diameter for a straight shaft pier ($d_b/d = 1$).
b. Maximum force in the pier.
c. Uplift movement of the pier.
d. End bearing load at the tip of the pier.

Solution.

Compute the free field heave, ρ_0, using Eq. (4.15) (see Example 5). Divide the active zone into five layers.

0 to 4 ft

$$\sigma_f' = 130 \times 2 = 260 \text{ psf}$$

$$\rho_1 = \frac{0.038 \times 4.0}{1.62} \log\left(\frac{260}{7400}\right) = -0.14 \text{ ft}$$

4 to 8 ft

$$\sigma_f' = 130 \times 6 = 780 \text{ psf}$$

$$\rho_2 = \frac{0.038 \times 4.0}{1.62} \log\left(\frac{780}{7400}\right) = -0.09 \text{ ft}$$

8 to 12 ft

$$\sigma_f' = 130 \times 10 = 1300 \text{ psf}$$

$$\rho_3 = \frac{0.038 \times 4.0}{1.62} \log\left(\frac{1300}{7400}\right) = -0.07 \text{ ft}$$

12 to 16 ft

$$\sigma_f' = 130 \times 14 = 1820 \text{ psf}$$

$$\rho_4 = \frac{0.038 \times 4.0}{1.62} \log\left(\frac{1820}{7400}\right) = -0.06 \text{ ft}$$

16 to 20 ft

$$\sigma_f' = 130 \times 18 = 2340 \text{ psf}$$

$$\rho_5 = \frac{0.038 \times 4.0}{1.62} \log\left(\frac{2340}{7400}\right) = -0.05 \text{ ft}$$

Then,

$$\rho_0 = \Sigma \rho_i = -0.41 \text{ ft} = -5 \text{ in.}$$

a. Length and diameter can be determined using Eq. (5.5b). Assume $d = 2.0$ ft for initial estimate.

$$L = 20 + \frac{1}{800}\left[0.16 \times 7400 \times 20 - \frac{1.0 \times 30{,}000}{\pi \times 2.0}\right] = 43.6 \text{ ft}$$

Try a 40-ft-long pier for subsequent design calculations.

b. P_{max} can be determined from Figure 5.10. Consider the pier-soil strength to be uniform with depth.

$$f_u = \alpha_1 \sigma_s' = 0.16 \times 7400 = 1184 \text{ psf}$$

From Eq. (5.10):

$$P_{FS} = -1184 \times \pi \times 2.0 \times 40 = -297 \text{ kips}$$

Assume $E_s = 100 f_u = 118{,}400$ psf.

$$\frac{\rho_0 E_s}{d f_u} = \frac{0.41 \times 118{,}400}{2.0 \times 1184} = 20.5$$

For $z_a/L = 0.5$, from Figure 5.10a:

$$\frac{P_{max}}{P_{FS}} = 0.36$$

and

$$P_{max} = 0.36 \times -297 = -107 \text{ kips}$$

The negative sign indicates tension in the pier.

c. Uplift movement of the pier can be determined from Figures 5.10b and 5.11. For no dead load (Figure 5.10b):

$$\frac{\rho_0}{d} = \frac{0.41}{2.0} = 0.2$$

Reading from Figure 5.10b:

$$\frac{\rho}{\rho_0} = 0.14$$

and

$$\rho = 0.14 \times 0.41 = 0.06 \text{ ft} = 0.69 \text{ in.}$$

The effect of dead load on the pier movement can be determined from Figure 5.11. For $L/d = 20$ and deep soil,

$$\frac{\rho_p E_s d}{P_{dl}} = 0.095$$

from which

$$\rho_p = \frac{0.095 \times 30,000}{118,400 \times 2.0} = 0.01 \text{ ft} = 0.14 \text{ in.}$$

The dead load will cause settlement whereas the expansion will cause heave. Therefore,

$$\rho_{total} = 0.69 - 0.14 = 0.55 \text{ in. (upward)}$$

d. End bearing on the pier can be determined from Figure 5.12. For $d_b/d = 1$, $v_s = 0.3$, and $L/d = 20$,

$$\frac{P_b}{P_{dl}} = 0.05$$

and

$$P_b = 0.05 \times 30,000 = 1500 \text{ lb}$$

5.2.1.3 Skin Friction

Innovative designs have been utilized in efforts to minimize uplift forces on piers in swelling soils. One system that has been utilized in Texas and Colorado consisted of a steel pipe placed inside the drilled hole and filled with concrete. Concrete was also placed outside of the pipe such that an annulus of concrete approximately 1 to 2 in. thick was created between the soil and the pipe. The concrete-filled pipe was designed to carry the structural loading. The outer perimeter of the pipe was coated with bitumen from the top down to the bottom of the active zone. The bitumen coating minimized the bonding between the pipe and the outer annulus of concrete forming the pier shaft. Because the concrete in the outside portion of the shaft was not reinforced, cracking should occur in the concrete at a point where uplift pressures exceeded the tensile strength. The broken outside concrete skin should move with the swelling soil and provide relief from uplift pressures on the supporting steel pipe. Although this technique appears to have been successful, care must be taken to assure that bonding does not occur between the concrete annulus and the pipe.

Other types of innovative techniques for reducing adhesion in the uplift zone around pier shafts include polyethylene sleeves, roofing felt, or PVC outer layers. Pervious material such as vermiculite or pea gravel should be avoided because they can provide a pathway for water to access deeper layers causing deep seated heave.

5.2.1.4 Construction Techniques and Quality Control

Drilled piers may be constructed in dry or cased holes, or by slurry displacement (Reese and Wright, 1977). Casing and slurry displacement techniques are used when the hole will not stand open prior to concrete placement. Concrete strength of at least 300 psi (20 MPa) is recommended, and the concrete should be cast immediately after the holes are drilled. The slump of the concrete should be 4 to 6 in. (100 to 150 mm) and the aggregate size should be limited to one-third of the reinforcement bar spacing to facilitate the flow of concrete around the reinforcement. Vibration of the concrete to help eliminate voids in the pier is important. Vibration by wedging the vibrator in the reinforcing steel should be prohibited.

A particular problem to be avoided is "mushrooming" of the pier near the top. This results if the upper edges of the hole are not formed properly, such that when concrete is cast the upper foot or so is much wider than the main portion of the shaft. The mushroom shape provides added area for uplift forces. The use of cylindrical cardboard forms at, or extended above the top of the concrete helps prevent formation of mushroomed piers.

Reinforcing steel should extend the entire length of the pier and should be hooked into the belled bottom and into the grade beam at the top. The area of steel should be designed to resist all tensile loads to which the pier may be subjected. Current practice is to use a minimum of 0.5% to 1% of the cross-sectional area of the pier.

The grade beam should be isolated from the swelling soil by construction of a void space beneath the beam between piers. The most certain way of providing the void space is by hand excavation after the grade beam has been cast. The provision of a crawl space has the advantage of allowing the void space to be maintained.

Commercial cardboard forms can be obtained that are capable of supporting concrete when it is intact. The forms are wrapped in plastic. When the plastic is punctured, the cardboard will deteriorate when wetted, thereby creating a void and preventing transmission of swell pressures to the grade beam. There have been instances, however, when these forms have not functioned as intended. If they do not get wet and deteriorate they can transmit significant uplift pressure to the beam. Obviously, it is not recommended to introduce water to provide for wetting. To avoid uncertainty, these forms can be manually removed after the beam has been cast.

Foamed plastic forms of polystyrene have been used in place of the void in some cases. They are not recommended, however, because they have a high crushing strength and can transmit significant uplift pressure to the grade beam.

The void space should be kept open using soil retainer planks and spacer blocks. The size of the void space depends on the predicted magnitude of potential swell of the soil beneath the beam. A minimum of 6 in. (150 mm) is recommended, and spaces as large as 12 in. (300 mm) may be needed.

When properly designed and constructed, drilled pier and beam systems can provide a reliable foundation system, with low risk, even in highly expansive soil areas.

5.2.2 Stiffened Slabs-on-Grade

5.2.2.1 General Description

Reinforced concrete slabs with a grid of underlying crossbeams as shown in Figure 5.13 have been used successfully as foundations for even relatively heavily loaded structures on expansive soils. Stiffened slab foundations are excellent foundation

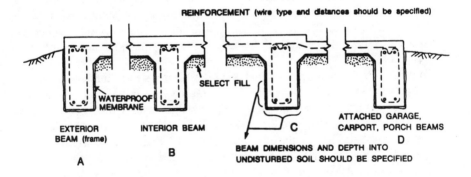

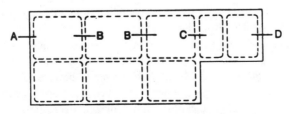

FIGURE 5.13. Typical detail of a reinforced slab on grade.

systems in areas where basements are not used or where expansive soil conditions extend to depths that make pier construction costs prohibitive. Stiffened slab construction is common in many areas in the southern and southwestern United States and in California. Stiffened slab foundations are also used extensively in Australia, Israel, and South Africa.

Stiffened slabs may be either conventionally reinforced or post tensioned. The design procedures basically consist of determining bending moment, shear, and deflection for applied structural loads and expected heave patterns. The input soil parameters include the amount of free field heave predicted, the heave edge distance (mound geometry), and the modulus of the soil. The same general method of analysis can be applied to the design of either conventionally reinforced slabs or to post tensioned slabs. Methodology for designing post tensioned slabs is available through the Post Tensioning Institute (PTI, 1980). Standard specifications and design methodology for conventionally reinforced slabs have not been published to date but individual design firms have developed their own methods based on the theory for beams on elastic foundations.

5.2.2.2 Design Considerations

Design of stiffened slab foundations on expansive soils is based on modeling the soil–structure interaction at the base of the slab. Theoretically, the slab and swelling soil are modeled as a loaded plate or beam resting on an elastic foundation. The general nature of the design principle is shown in two-dimensional form in Figure 5.14. Figure 5.14a shows the mound that would occur beneath a weightless slab with no load applied. The mound is characterized by the maximum heave at no applied effective stress. This value of ρ_{max} would correspond to the free field heave determined by Eq. (4.23), and will be equal to the maximum mound height, y_{max} (Figure 5.14a).

Figure 5.14b shows the mound that would result if a slab of infinite stiffness is placed on the mound with the applied structural loading. In this case, the mound is characterized by the maximum heave, y_m, which is smaller than y_{max} in Figure 5.14a. This would correspond to the heave as predicted by Eq. (4.16). The distance, E, which is the distance from the outer edge of the slab to the point where it contacts the mound is also necessary to characterize the mound in Figure 5.14b.

The situations depicted in Figures 5.14a and b represent the two extremes. In reality, the slab will have some flexibility and the actual mound shape will be intermediate between those shown in Figure 5.14a and b. The actual situation is shown in Figure 5.14c.

Figure 5.15 shows the load, shear, and bending moment diagrams along the axis of the slab in Figures 5.14b and c.

The design procedure consists of predicting the mound shape and height and the relative stiffnesses of the soil and the slab. Consideration of the soil–structure interaction allows the shear and bending moments to be determined.

Several different design approaches have been developed, each prescribing a different combination of soil and structural design parameters. Five theoretical design procedures for stiffened slabs-on-grade are summarized in Table 5.6. Others exist as well but are not included here. These five procedures will be discussed

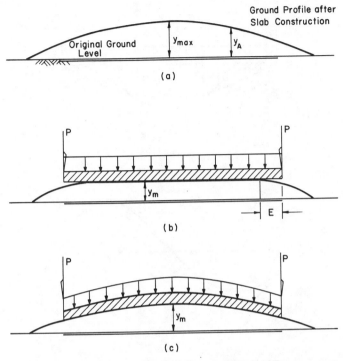

FIGURE 5.14. Mound profiles for various cases of slabs on grade: (a) mound profile after slab placement if no load is applied, i.e., free field heave with weightless slab (Case A); (b) mound profile for infinitely stiff slab with load (Case B); (c) mound profile for flexible/semirigid slab with load (Case C).

below with respect to the methodology employed. Before that, however, the general input parameters will be discussed.

The primary geotechnical task in slab-on-grade design is to define the size, shape, and properties of the distorted soil surface, or "mound" that may develop beneath the slab. The design parameters used to define the mound characteristics vary somewhat for the different design procedures followed. As shown in Table 5.6, these mound characteristic design parameters include the following:

- Maximum differential heave (y_m)
- Edge "moisture variation" distance (e)
- Support index (C)
- Mound shape equation constants (m, a)
- Climatic ratings, (C_w, TMI)
- Soil properties (k, q_a, E, u, f)

The shape of the soil surface that will develop beneath a slab depends on heave, soil stiffness, initial moisture conditions, moisture distribution, climate, elapsed

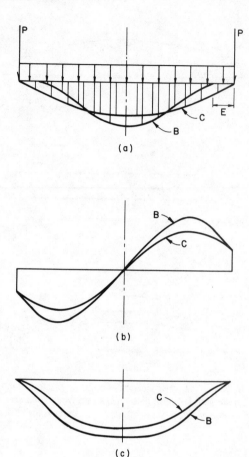

FIGURE 5.15. (a) Load, (b) shear, and (c) bending moment distributions for stiffened slabs on grade.

time from construction, structural loads, rigidity of slab, and many other variables that are difficult to quantify.

The preceding discussion has considered a mound shaped deformation pattern. Also possible is a dish shaped pattern as shown in Figure 5.16. These two patterns are generally referred to as center heave or edge heave conditions.

Edge heave results when the exterior of the structure increases in water content before the interior areas do. This would be the case, for example, with irrigation of landscaping around buildings.

Center heave results most commonly from increased water content as a result of elimination of evapotranspiration as shown in Figure 2.2. The increase in water content is greatest near the center of the structure and evapotranspirative losses around the edge cause a lesser degree of heave near the edges. Center heave represents the long-term most severe distortion condition (Lytton, 1970). Assumptions

TABLE 5.6. Summary of five stiffened slab-on-grade foundation design methods[a]

Design Method	1 BRAB (1968)	2 Lytton (1972)	3 Walsh (1978)	4 Swinburne (Holland et al., 1980)	5 PTI (1980)
Slab load and soil–structure interpretation assumptions	RIGID MOUND	WINKLER k	COUPLED WINKLER k PARABOLIC EDGES	ELASTIC MOUND E, μ PARABOLIC EDGES	ELASTIC MOUND E, μ CUBIC EDGES
Mound shape assumptions	Rectangular mound. Empirical support index (c) related to climatic rating and soil properties	Parabolic mound $y = ax^m$	Flat under center to a distance of E from edges. Parabolic at edges	Flat under center. Parabolic at edges	Flat under center. Cubic at edges
Design parameters					
1. Climate	1. C_w	1. (N/A)	1. (N/A)	1. (N/A) E_s, ν	1. TMI (Figure 2.5)
2. Soil parameters	2. C	2. y_m	2. y_m, e, k	2. y_m, e	2. y_m, e, q_a, f
3. Loading conditions	3. w	3. w, q_c, q_e	3. w	3. w, I_c, q_c, q_e	3. w, q_e, q_c
4. Slab parameters	4. L, E	4. L, S, E	4. L, E	4. L, b, S, E	4. L, b, d, S, E

[a]Legend: a = mound equation constant; b = beam width; C = support index (Figure 5.18); C_w = climatic rating factor (Figure 2.6); d = beam depth; e = edge distance; E = soil modulus of elasticity; E_s = subgrade modulus; L = slab length; m = mound exponent; ν = Poisson's ratio of soil; w = average foundation pressure; q_a = allowable soil bearing pressure; q_c = center load; q_e = edge load; S = beam spacing; f = slab-subgrade friction coefficient; E = edge distance.

149

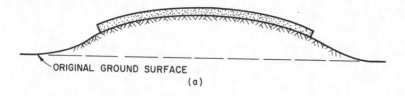

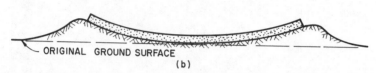

FIGURE 5.16. Two principal mound shape conditions: (a) center heave and (b) edge heave (Walsh, 1978, courtesy of CSIRO Australia).

for various center heave mound shapes are summarized in Table 5.6. The mound size and shape are defined in most design procedures by the maximum differential heave (y_m), the edge "moisture variation" distance (e), and a mound shape factor (m).

Maximum differential heave (y_m) is calculated as a selected percentage of the maximum total predicted heave (ρ_{max}). Heave prediction may be performed using any preferred technique, but the oedometer and soil suction procedures described by Eqs. (4.16) and (4.23) are recommended. Observations of total and differential heave of slabs at various sites indicate that y_m varies from 33 to 100% of the total maximum heave (e.g., Holland and Lawrence, 1980; Poor, 1978; Donaldson, 1973).

The edge distance, or edge moisture variation distance (e) has been defined as "the distance measured inward from the slab edge over which the soil moisture varies enough to cause soil movement" (Wray, 1980). However, as indicated in Figure 5.14 there actually will be some mound development and, hence, soil movement beginning at the edge of the slab. This edge distance should be a function of the distance, E, as shown in Figure 5.14, climatic conditions, relative stiffnesses of the slab and soil, boundary conditions of the slab, and other factors. For any definition it is evident that the edge distance is nebulous to define and the most difficult design parameter to estimate. From measurements that have been made, it appears that this distance typically ranges from 2 to 6 ft (0.6 to 1.5 m) (Wray, 1980). McKeen and Johnson (1990) indicate that it can be much larger, approaching the distance of the depth of the active zone.

It is implied in its definition that for the center heave case, the edge distance is defined with respect to the long-term conditions that exist after an "equilibrium" moisture condition has developed beneath the center of the slab. The edge distance is the distance inward around the slab edge that is most subject to seasonal drying or wetting with respect to the "equilibrium" center moisture condition.

The PTI (1980) method recommends a procedure for estimating edge distance based on the Thornthwaite Moisture Index (TMI). The TMI was defined in Chapter 2, and the TMI distribution for the United States is shown in Figure 2.5. Figure 5.17 shows approximate relationships between TMI and the edge distance for both the center heave and edge heave cases of mound development (Wray, 1978). This figure is probably the most widely used data for estimating edge distance.

The support index (C) is an empirical parameter related to the elastic modulus of the soil, used in the BRAB design procedure to account for the fact that the soil heave will be decreased by the load applied by the slab. It is used as a multiplier to adjust the calculated bending moments. The BRAB support index is estimated based on the edge distance (E), soil properties, and the BRAB climatic factor (C_w), as shown in Figure 5.18. The BRAB climatic factor was defined in Chapter 2 and is shown in Figure 2.6.

The mound exponent (m) defines the curvature of the mound. The mount exponent has been related to the ratio of the length of the slab to the depth of the active zone (Lytton, 1972; Mitchell, 1980). By assuming that the mound is flat beneath the interior of the slab, the mound exponent will seldom exceed 7 or 8 (Walsh, 1978). A mound exponent of 2 provides the least support and is the most conservative condition for analysis purposes. Mound exponents equal to 3 or 4 are commonly used for slab design in Australia (Woodburn, 1974).

The soil properties used in the various structural analyses to characterize the soil–structure interaction include allowable soil bearing pressure, subgrade modulus, soil modulus of elasticity, Poisson's ratio of the soil, and the slab-subgrade friction coefficient.

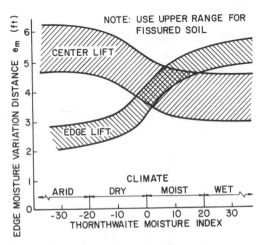

FIGURE 5.17. Approximate relationship between Thornthwaite Moisture Index and edge moisture variation distance (Wray, 1978).

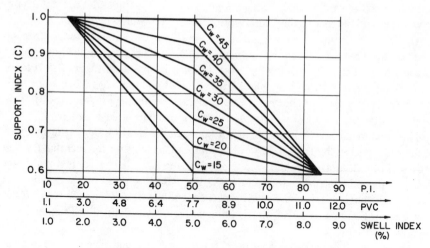

FIGURE 5.18. BRAB Support Index (C) (BRAB, 1968).

5.2.2.3 Design Procedures

All slab design methods summarized in Table 5.6 use the procedure of dividing an irregularly shaped slab plan into overlapping rectangles (Figure 5.19) and designing each rectangle independently. In addition, the design methods generally adopt a value of 4 to 6 in. (100 to 150 mm) for slab thickness.

With the exception of the Walsh method, all procedures calculate the design moment, shear, and deflection. The calculated deflection is then compared with the allowable structural deflection.

The various design methods are described briefly below. Refer back to the original reference or a detailed user manual to use a particular method for an actual design would require reference.

5.2.2.3.1 Building Research Advisory Board (BRAB) Procedure. The first step in the BRAB procedure is to determine the slab type needed to accommodate the soil and climatic conditions. BRAB separated slabs into four types as follows:

Type I   Unreinforced
Type II   Lightly reinforced against shrinkage and temperature cracking

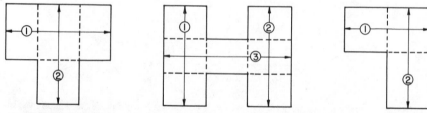

FIGURE 5.19. Typical design rectangles for slabs of irregular shape (BRAB, 1968).

Type III   Reinforced and stiffened
Type IV   Structural (not directly supported on the ground)

Slab type III is typically used for expansive soils. The principal assumptions for designing slab type III are as follows:

1. The total superstructure load is uniformly distributed over the entire slab area.
2. The soil pressure is distributed evenly beneath the supported slab area.
3. The value of the support index (C) (Figure 5.18) is indpendent of slab dimensions.
4. For analysis of deflection and stress distribution, each rectangular slab is analyzed as two, one-dimensional cases rather than a two-dimensional analysis.

Once the support index is determined, based on Figure 5.18 and the climatic rating map (Figure 2.6), the values of maximum moment, shear, and deflection are calculated.

The maximum differential deflection can then be compared with the allowable value for various superstructures as listed in Table 5.7.

5.2.2.3.2 *Lytton Procedure.*

Lytton's design methods (Lytton, 1970, 1972, 1973; Lytton and Woodburn, 1973) were developed with closed form solutions, except for the 1972 procedure, which used a finite difference analysis. The approach for calculating the maximum moment for each direction of a rectangle is based on the assumption that both soil and footing are rigid and the soil provides only line support. He assumed three types of loads acting on the slab: line loads acting along the slab edge (q_e), line load acting through the center of the slab (q_c), and a uniformly distributed load (w) resulting from interior dead and live loads. The soil support index in Lytton's method is based on the subgrade modulus (k) and the mound shape.

The maximum moment for center heave was calculated for the structural loading and soil reaction. The calculated maximum moment was then reduced to account for the compressibility of the soil and taking into account factors determined by two-dimensional slab analysis.

With the design moments, the dimensions of edge and internal beams can be calculated. The Lytton procedure is usually limited to slabs with a maximum dimension of 85 ft (26 m) (Lytton and Woodburn, 1973).

5.2.2.3.3 *Walsh Procedure.*

The design procedure proposed by Walsh (1978) is based on analysis of a beam on an elastic coupled Winkler foundation. Based on a parametric study of soil and structural variables, Walsh expressed the beam-on-mound problem in terms of three nondimensional parameters that he correlated with two support indices. The support indices were tabulated for various combinations of the nondimensional parameters.

From trial calculations, Walsh concluded that the shear strength of the slab was not an important design consideration. He provided equations for design moment and deflection.

TABLE 5.7. Allowable differential deflection ratios for slab-on-grade to limit damage to superstructure

Type of Construction	Maximum Allowable Deflection Ratio (Δ/L)	Reference
Wood frame	1/200	BRAB (1968)
Nonmasonry, timber or prefabricated	1/200	Woodburn (1974)
Unplastered masonry or gypsum wallboard	1/300	BRAB (1968)
Nonmasonry, frame, and panel	1/300	Mitchell (1980)
Stucco or plaster	1/360	BRAB (1968)
Brick veneer (articulated)	1/300	Woodburn (19740
Brick veneer	1/480	Walsh (1978)
Brick veneer (standard	1/500	Woodburn (1974)
Masonry completely articulated)	1/500	Mitchell (1980)
Masonry (partially articulated)	1/800	Woodburn (1974)
Fully articulated solid brick	1/1000	Holland and Lawrence (1980)
Masonry, solid or cavity wall	1/1500–1/2000	Woodburn (1974); Holland (1980)

Modified from Wray (1978).

The Walsh design procedure can be summarized as follows:

1. The soil parameters of maximum differential heave (y_m), edge distance (e_m), and subgrade modulus (k) are determined. The load per unit length (w) is calculated. The allowable displacement is selected from Table 5.7 according to the selected slab span (L) and the type of structure.
2. The values of the soil support indices are determined based on the selected values of the dimensionless parameters using the tables presented in Walsh (1978).
3. The design moment and structural stiffness are calculated by equations presented in Walsh (1978).

Walsh also provided charts for selection of beam dimensions for center heave conditions.

5.2.2.3.4 Swinburne Procedure. The Swinburne method is based on work conducted by Australian researchers Fraser and Wardle (1975). A finite element method of analysis is used to model the slab as a plate resting on a semiinfinite elastic soil. The Swinburne method uses design charts to calculate moment, deflection, and beam depth for assumed values of maximum differential heave (y_m), edge distance (e), concrete strength, and the number and width of underlying cross beams. A complete description and example of the procedure is presented in Holland et al. (1980).

5.2.2.3.5 Post Tensioning Institute Procedure. The PTI (1980) design method was based on research conducted at Texas A&M University by Wray (1978). It is fairly widely used and is recognized in the Uniform Building Code. Modeling the soil–structure interaction as a plate resting on an elastic continuum, Wray analyzed various sets of input variables (y_m, e, beam spacing and depth, loading, and slab length) using finite element techniques. Once all the soil–structure interaction data were accumulated, a regression analysis was performed to produce general design equations for center heave and edge heave conditions. The general equations for design moment, shear, and deflection are presented in PTI (1980) and Wray (1978, 1980), as functions of the above soil and structural variables. The PTI procedure consists of the following steps:

1. Assemble all the soil and structural data needed for design.
2. Assume a trial section in both the long and short directions of each rectangular design section.
3. Calculate the expected moment and shear forces of the assumed section in each direction.
4. Determine the magnitude of allowable moment and shear forces and compare with the calculated values.
5. Determine if the trial section will meet differential deflection criteria in each direction.

Wray indicated that the design procedure must be conducted for conditions of both center and edge heave to ensure that the section is adequate in either distortion mode.

5.2.2.4 *Construction Techniques and Quality Control*

Figures 5.20 and 5.21 show the construction of a stiffened slab-on-grade foundation.

(a)

(b)

FIGURE 5.20. Construction of a stiffened slab on grade for a large commercial building showing (a) cleanout and inspection of beam trenches and (b) a "cold joint" in the beam (courtesy U.S. Army Corps of Engineers, Fort Worth District).

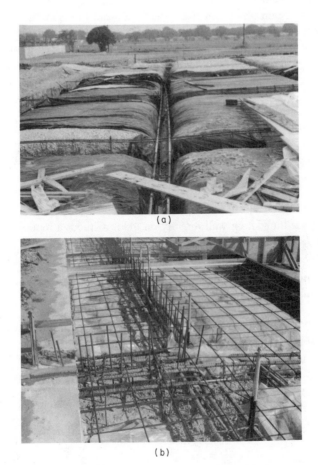

FIGURE 5.21. Construction of a stiffened slab on grade showing (a) placement of waterproof membranes and (b) slab reinforcement (courtesy U.S. Army Corps of Engineers, Fort Worth District).

Stiffened slab-on-grade construction should conform to the requirements of the American Concrete Institute (ACI) Standard 302-69 "Recommended Practice for Concrete Floor and Slab Construction" (American Concrete Institute, 1969). Construction joints should be placed at intervals of less than 150 ft (46 m) and cold joints at less than 65 ft (20 m).

Post tensioned slabs require trained personnel and careful inspection to properly apply the post tensioning procedure. Specifications for placing the post tensioning tendons and conventional reinforcement are provided by the Post Tensioning Institute (PTI, 1980).

Waterproof membranes or moisture barriers may be placed over the subgrade prior to slab or beam placement. Care must be taken to assure that the membrane does not become torn or entangled with the reinforcing steel. Indications are that plastic-type waterproof membranes generally do not exhibit long-term integrity and are useful only in the short term to minimize wetting of the subgrade.

5.2.3 Shallow Footing Foundations

Shallow spread footings typically are not used in expansive soil applications. Where shallow footings are used, techniques usually are applied in an attempt to increase the bearing pressure so as to minimize heave. Some modifications that have been used include

- narrowing the width of the footing base
- placing the foundation wall directly on grade without a footing
- providing void spaces within the supporting beam or wall to concentrate loads at isolated points
- increasing the reinforcement around the perimeter and into the floor slab to stiffen the foundation.

Figure 5.22 shows one type of modified continuous perimeter wall footing in common use in areas with low to moderate swell potential. The slab and footing are cast as a unit and reinforced to resist rotational pressures on the footing.

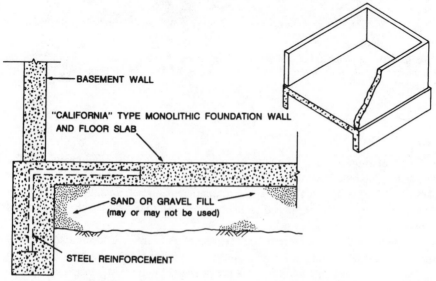

FIGURE 5.22. "California" type continuous perimeter wall footing.

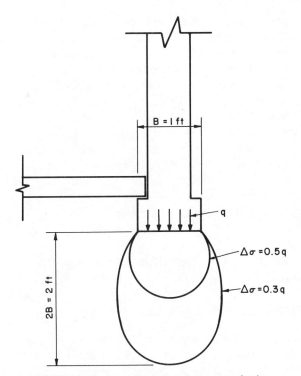

FIGURE 5.23. Pressure bulb below narrow footing.

Shallow wall footings or narrow spread footings may be used where the top layers of expansive soil are thin and a stable nonexpansive stratum can be reached on which the footings can bear. In some cases a narrow footing width, generally less than 12 in. (300 mm), has been used to increase the bearing stress to minimize heave. However, the pressure bulb within which the increased stress is sufficient to reduce heave extends only to a depth of approximately two footing widths. For narrow footings this depth is relatively small. The pressure bulb beneath a narrow footing is shown in Figure 5.23. If the load on the footing is 10,000 psf the applied stress at a depth of 1 ft would be 2500 psf and at a depth of 2 ft would be 1000 psf. Thus, even if the swelling pressure of the soil is as low as 1000 psf only the soil to a depth of 2 ft would be prevented from swelling. This depth is much smaller than that of the active zone for most sites. This footing would be ineffective in preventing heave.

It is evident that the use of narrow spread footings in expansive soils should be restricted to soils having less than 1% swell potential and very low swell pressures.

"Box" construction has been used in cases where basements are included. In this technique the foundation walls form the "box" and are heavily reinforced so that they can span unsupported distances and resist cracking due to differential movement. Two No. 5 bars are typically placed at both the top and bottom of the wall at a minimum of 30 in. (76 cm) vertical spacing (Sealy, 1973). The structural configurations should be simple, and for split level construction the structure should be designed in terms of independently acting units connected by joints. Masonry bricks and cinder blocks should not be used without reinforcement.

Another method used to concentrate loads and increase bearing stress is to place shallow individual footings at several points under the foundation walls. This system has the same limitation as shown in Figure 5.23, and has only limited success at decreasing heave. Therefore, the superstructure should be relatively flexible. Load concentration on the pads is accomplished by providing a void space beneath the grade beam using the same technique as described for pier and grade beam construction. It is important to make sure that uplift pressures on the sides of the pad are minimized or prevented. Care must be taken to assume that in concentrating the loads on the pads, the bearing capacity of the soil is not exceeded.

5.3 MOISTURE CONTROL AND SOIL STABILIZATION FOR FOUNDATIONS

5.3.1 Moisture Control Methods for Foundations

Because the source of swelling derives from an increase in water content of the expansive foundation subsoils, one obvious method for controlling swell would be to control the moisture. However, it is virtually impossible to prevent an increase in water content of the foundation soils. It is possible to control the rate of increase and minimize seasonal fluctuations.

Research at Colorado State University by Porter (1977), Nelson and Edgar (1978), and Goode (1982) has shown that even if a distinct water table does not exist in the immediate depth below the zone of seasonal fluctuation, an increase in water content in the upper 10 to 20 ft (3 to 6 m) of soil will occur when evapotranspiration is eliminated. This has been discussed at several points previously (i.e., Figure 2.2). Generally, the increase in water content within the upper zone takes place over a period of time ranging from only a few years up to 10 years.

In view of the futility of avoiding the increase in water content within foundation subsoils, moisture control can only be expected to stabilize water contents over time and minimize fluctuations. This can be effective in minimizing damage to structures and pavements if the initial heave can be kept relatively uniform and slow.

It has been common practice in some areas to recommend the use of drains around foundations so as to avoid the presence of free water in the subsoil. Generally, for expansive clays, soil suction is high and permeability is very low. Consequently, perimeter drains are virtually useless in avoiding heave below foundations. In fact, the presence of the drain may provide for a source of water if long-term positive drainage is not ensured.

An innovative scheme has been developed by Schmertmann (private communication) in Gainesville, Florida to provide for the stabilization of water contents.

Problems in that area relate to shrinkage arising from decreases in water content by evapotranspiration. In Schmertmann's scheme, small-diameter wells are drilled at intervals around the foundation. A header pipe is connected to each of these wells and it is connected to a sump with a controlled water level. This provides a constant low head water supply to the soil. It is a low maintenance system and has been shown to be effective in stabilizing water contents in relatively wet climates. This method appears to work well in Florida because the high water tables also keep the interior moist. However, it has not had good success in dry climates. This procedure is discussed in detail in Section 7.2.2.8.

Another scheme for controlling and stabilizing water contents beneath slabs has been the utilization of horizontal and vertical barriers. Horizontal barriers are membranes that extend outward for considerable distances around the edges of the foundation or floor slab.

Examples of large horizontal barriers are parking lots around convenience stores. Although these aprons do not eliminate an increase in water content, for reasons discussed above, they do provide for a more uniform heave pattern to develop around the foundation or floor slab of the building. Because of the large extent of these barriers the edge effects are removed to a considerable distance away from the structure.

Another scheme to stabilize water contents has been the utilization of vertical barriers. Goode (1982) placed vertical moisture barriers to depths as great as 8 ft (2.5 m) around simulated floor slabs placed on natural ground. Measurements were made of the surface heave and water content profiles over a period of several years. The results indicated that, as for horizontal barriers, increases in water content still occurred and the amount of heave was not decreased. However, the heave was significantly more uniform for the slabs with vertical barriers than for slabs without vertical barriers. Also, seasonal fluctuations were less. This is due primarily to the reduction in edge effects by the barriers. Goode's (1982) experiments are discussed in some detail in regard to vertical moisture barriers in Chapter 6.

5.3.2 Soil Stabilization

Techniques are also available in which the characteristics of the expansive soils can be altered or the soil can be removed and replaced. These methods can be used alone or in conjunction with specific design alternatives. They are discussed briefly at this point with more details being given in Chapter 6.

5.3.2.1 *Removal and Replacement*

In this method the problem soils are removed to some depth and replaced with a compacted nonexpansive fill. Factors that need to be considered are depth of removal, and the amount, location, and cost of the fill. Laboratory tests and engineering judgment can be used to determine the depth of the active zone and the potential uplift. This information can be used to determine the necessary depth of soil to be removed. Even if the entire depth of expansive soil is not removed, placement of a structure on compacted nonexpansive fill underlain by expansive soils will produce a more uniform heave pattern, resulting in less structural damage from differential heave. The maximum practical depth of removal is about 4 ft (1.5 m). If nonexpansive fill is not available the swell characteristics of the expansive soils on site can be altered by excavating and recompacting using compaction control. Compacting the material wet of optimum can decrease the swell potential. Also, compaction to a relatively low density will minimize heave, but care must be taken to assure that at the lower density the soil has adequate strength and compressibility.

5.3.2.2 Prewetting

Ponding or prewetting the soil can be used to cause heave to occur prior to construction. Although this method has been used with some success on a few projects, a large number of potential problems can be associated with the prewetting technique. If the soil has low permeability, as do expansive clay shales, excessive time will be required for wetting. The soil mass and the depth of wetting will be very limited. Furthermore the soil can undergo a serious reduction in bearing capacity as the soil becomes saturated. Wetting of deep expansive layers can continue over long periods of time causing future heave.

Fissured and fractured soils respond more favorably to prewetting because of access paths for the water. Also a grid of sand drains can be installed to increase the wetting surface and decrease the time factor. In some areas of southern California this method has met with some success when used in soils of low expansivity and relatively high permeability.

5.3.2.3 Chemical Stabilization

The most common chemical admixtures used in soil stabilization are lime and cement. There are other organic and inorganic compounds available, but generally they are not economically viable.

If soils are lime reactive, the addition of 2 to 8% lime can reduce the plasticity and swell potential of the soil and increase its shrinkage limit. The soil texture can be altered to improve its workability. Hydration reactions will increase the shear strength of the soil. The most effective method of adding the lime is mixing in place.

The clay is excavated to a predetermined depth and broken into clods 4 to 6 in. (10 to 15 cm) in size. Lime can be applied dry, with water added to the soil–lime mixture, or in slurry form. Quality control is important, and the effectiveness of the treatment is limited by the thoroughness of the mixing procedure.

Pressure injection of lime is a technique that produces varying results. Lime migration is a very slow process in a soil mass, and is not significantly increased using pressure injection. In cases where extensive fissuring exists in the soil, the technique has reportedly had success. One possible explanation is that the lime in the fissures helps in stabilizing the water content in the soil between the fissures.

Cement stabilization is similar to that of lime and produces similar results. These include a reduction of the liquid limit, plasticity index, and potential volume change and an increase in the shrinkage limit and shear strength. Cement-treated soils can exhibit higher strength gains than lime-treated soils.

5.4 DESIGN APPROACHES AND TREATMENT ALTERNATIVES FOR HIGHWAY AND AIRFIELD PAVEMENTS

At least one-half of the total dollar losses attributed to expansive soil movements is due to pavement damages. Pavements are particularly susceptible to damage by expansive soils because they are lightweight and extend over large areas. As a rule, it is not economically viable to bypass expansive soil areas or to remove the problem materials and replace them by other, more stable soils in highway construction.

Worldwide experience has shown that pavements on expansive soils often require costly rehabilitation well before the end of their design life. Damage to pavements on expansive clays appears in four major forms (Kassiff et al., 1969):

- Severe unevenness along a significant length of the pavement with or without cracking or visible damage.
- Longitudinal cracking, parallel to the pavement centerline.
- Significant localized deformation, for example, near culverts. This is generally accompanied by lateral cracking.
- Localized failure of the pavement, associated with disintegration of the surface.

Pavement unevenness (first item above) is particularly important in airfield runway pavements. When aircraft travel across an uneven runway surface, unacceptable vertical accelerations may be developed. If these accelerations become too large, pilots are unable to safely negotiate the aircraft on the pavement, passengers are uncomfortable, and damage to the aircraft or cargo may occur (McKeen, 1981).

Pavement design for expansive soils is similar to that for foundations in that it must provide for safety against failure or excessive deformation and must be economical. However, the design strategy is different. Pavements cannot be isolated from the soil, and it is generally uneconomical to make them stiff enough to resist differential movements. Therefore, subgrade soil treatments are commonly employed to stabilize or minimize the soil movements. Commonly used subgrade treatments include mixing with lime or other stabilizing additives, injection of lime slurries, construction of moisture barriers, and control of placement density and moisture content of base course and compacted subgrade materials. Increased pavement and subbase thicknesses and careful control of grading are also employed. Although specific design criteria are not within the scope of this book, general pavement design principles are briefly described below, primarily for orientation to the discussion of design and treatment practices used for expansive subgrades.

5.4.1 General Principles of Pavement Design

Pavement systems are classified as either rigid or flexible pavements for design purposes. Rigid pavements are constructed of Portland cement concrete (PCC) and are designed primarily on the basis of the structural strength of the concrete. Theoretical analyses for rigid pavements such as the Westergaard (1927) approach express the relative stiffness of the slab and subgrade. It is assumed in the analysis that the deformation of the concrete-subgrade system may be modeled as a beam supported continuously on an elastic foundation. The design of the pavement thickness is based on the flexural strength of the concrete. The subgrade is generally characterized by the modulus of subgrade reaction (k) that is analogous to the spring constant of an elastic foundation.

Rigorous theoretical design analysis is typically used only for airfield pavement design. The required thickness is adjusted by a factor of safety which varies for the type of feature, i.e., taxiway, runway, or apron. The principal design variable for rigid highway pavements is number of repetitions of load plus the thickness of the base course to prevent pumping and frost action. Rigorous analysis is rarely performed for design of PCC highway pavements. In fact, most states in the United States use a standard cross section for rigid highway pavements (Yoder and Witczak, 1975).

A flexible pavement system typically consists of a relatively thin asphalt concrete surface layer built over thicker layers of granular base and subbase courses. The objective in flexible pavement design is to reduce the subgrade stress or deflection to a predetermined level, based on relevant failure criteria, by designing a system that is built up from the subgrade in layers having successively higher modulus values. A layer must not be so stiff, however, that excessively high tensile stresses would develop at the bottom of the layer or that high horizontal shearing stresses would develop between layers. The theoretical analysis for conventional flexible pavement systems utilizes Boussinesq type stress distributions in multilayered elastic materials.

Design criteria for flexible pavements are based primarily on subgrade strength. Although stiffer upper layers reduce subgrade stress and deflection, the presence of excessively stiff layers also induces high tensile stresses and increased horizontal shearing stresses. If an asphalt pavement has high stiffness, it may behave essentially as a rigid pavement, and conventional flexible pavement design criteria are not applicable. For example, a full-depth asphalt pavement, which is a pavement having asphalt mixtures in all courses above the subgrade, is essentially a rigid structure. Full-depth asphalt pavements are designed on the basis of the vertical compressive strain at the top of the subgrade (permanent deformation criteria) and the horizontal tensile strain at the bottom of the asphalt concrete layers (fatigue criteria). Thus, the thickness design may be dictated by either the subgrade strength if the asphalt concrete has low stiffness, or the fatigue resistance of the pavement layers if the asphalt concrete has high stiffness (Witczak, 1972).

Pavement design decisions for both flexible and rigid pavements are based on functional failure rather than structural failure. Structural failure would include the collapse or breakdown of one or more of the pavement components (Yoder and Witczak, 1975). On the other hand, functional failure implies that the pavement does not function as it was intended, that is, without causing high stresses or vibrations in the vehicles passing over it. Functional failure depends primarily on the degree of surface roughness, and may or may not be associated with structural failure.

5.4.2 Design Features and Treatment Methods for Expansive Pavement Subgrades

5.4.2.1 Highway Pavements

A large number of states in the United States utilize special design features and subgrade treatments for expansive soil areas. Design criteria are typically based on soil classification. Both the Unified Soil Classification System and the AASHTO "Group Index" classification procedures are used on a routine basis. Some states have developed empirically based design criteria using specialized tests. These tests include potential vertical rise (PVR), expansion index (EI), and stabilometer (R-value) tests.

Subgrade treatments and pavement design procedures are also fairly standardized in most state highway procedures. The practical treatment and design alternatives for highway construction may be separated into the following broad categories (Krazynski, 1980):

Choose alternative location. This may be done to avoid having to deal with expansive soils, but generally such an approach is not practical for highway use because the problem soils tend to occur over broad areas. Also, highway alignments are strongly influenced by other considerations. Nevertheless, if the alignment can be adjusted problems may be mitigated by such approaches as minimizing cuts and areas of poor drainage. The judicious choice of alignment can minimize the severity of the problem, if good reconnaissance surveys are made.

Removal and replacement. Removal of problem materials and replacement with nonexpansive soils can be useful. The nonexpansive soil produces a surcharge loading on underlying expansive soils as well as eliminating the expansive nature near the surface where moisture fluctuations are greatest. Generally, however, this approach is too costly because of typically long haul distances for nonexpansive soil material.

Physical alteration. In-place expansive soils can be ripped and scarified to destroy the natural structure of the materials and can be subsequently recompacted with good moisture and density control to minimize the expansion potential. This method is well known and widely practiced. In some instances sand has been added to the soil prior to recompaction to decrease swell potential.

Chemical alteration. Lime is the most effective and widely used chemical additive for expansive soils. It has been demonstrated to be effective when thoroughly mixed with pulverized clay materials in percentages ranging from about 3 to 6%. The depth of treatment is generally limited to about 8 to 12 in. (20 to 30 cm) in a single lift. Deep plowing techniques have been used to extend this depth to 2 ft (0.6 m) or more. The methodology for the use of lime is well established and numerous excellent results have been achieved.

Minimize moisture changes. Most highway pavements constitute a rather effective impervious membrane that reduces evaporation. As discussed with regard to foundations the elimination of evapotranspiration from the ground surface will produce an increase in water content. For practical purposes it is virtually impossible to maintain expansive subgrade soils in a low water content state. It may be possible to delay capillary rise of soil water, but highways generally have long design lives. Consequently, with evaporation minimized over a long period of time, wetting will almost certainly take place. Some measures that have been used to protect subgrade soil from excessive moisture fluctuation include

- Presaturation (ponding) with and without drill holes
- Vertical and horizontal membranes
- Deep sand drains
- Pressure injection of lime slurry.

Some of the design criteria, subgrade treatments, and pavement cross sections that have been used by U.S. State Highway departments for expansive soils were reported by the U.S. Army Engineer Waterways Experiment Station (WES) for the Federal Highway Administration (FHWA). The results of the WES state practice review are summarized in Table 5.8 (Snethen et al., 1975).

5.4.2.2 Experimental Highway Test Sections
Several state highway departments and other agencies have implemented experimental design concepts in test sections of roadway. Some of the major research projects are described below.

5.4.2.2.1 The Waco Ponding Project (Texas State Department of Highways and Public Transportation, TSDHPT). This project included 18 test sections along 8 mi (13 km) of southbound lands of Interstate Highway 35 in McLennan County, Texas. Areas to be ponded were selected on the basis of PVR values in excess of 1 in. (25 mm), using the TSDHPT procedure (Texas State Department of Highways and Public Transportation, 1970). Specifications included standard procedures for grading, drainage, and pavement construction, with the exception that selected areas of the natural ground and subgrade were ponded for up to 90 days. The structural section called for 6 in. (1950 mm) of lime stabilized subgrade, 5 in. (130 mm) of gravel basecourse, and 12 in. (300 mm) of nonreinforced concrete pavement. The results of the study indicated that ponding in general was beneficial and that the combination of ponding and lime stabilization of subgrade was successful in preventing pavement roughness due to volume changes in the subgrade.

5.4.2.2.2 San Antonio Deep Vertical Moisture Barriers (Steinberg, 1981). This project involved two test sections of interstate highway on highly expansive black clays. The first section was constructed in 1979 along a 0.5-mi (0.8-km) stretch of Interstate Highway Loop 410. This is a four-lane divided highway constructed on 16 in. (400 mm) of granular foundation course, and 8 in. (200 mm) of flexible base, with 3 in. (75 mm) of Type A and 2 in. (50 mm) of Type C hot-mix asphaltic concrete. As shown in Figure 5.24 a geotextile impermeable membrane was installed

TABLE 5.8. Some U.S. State Highway Department procedures for controlling expansive subgrades

State	Design Criteria	Subgrade Treatment	Special Design Features
Arizona			Full-depth asphalt
California	AASHTO (PCC pavements only) Expansion index tests Linear expansion tests R-values	Bituminous membranes Compaction moisture control 6–12 in. (15–30 cm) lime treatment in some districts	Reduce PCC thickness over lime stabilized subgrades No special design for flexible pavements
Colorado	AASHTO PI (moisture density control only)	Moisture-density control Bituminus membranes Drilled hole lime treatment (used infrequently)	Full-depth asphalt base with lined ditches Conventional flexible pavements with untreated bases use membrane lined subgrades and ditches
Kansas	Oedometer swell tests—remolded samples under 1 psi (6.9 kPa) surcharge to indicate swell potential	Subgraded up to 2 ft (60 cm) Lime treatment of uper 6 in. (15 cm), 5% Compaction moisture control	Conventional pavement sections Subsurface interceptor drains in cut sections
Louisiana	Plasticity index	Subgrades up to 3 ft (1 m) Compaction moisture 2% above optimum If moisture too high, treat full subgrade depth with 3% lime	Strict ditch depth control Conventional pavement sections
Mississippi	AASHTO	Lime treatment, 4 to 6% Subexcavate and replace in cut sections	Conventional pavement sections Use nonexpansive material in upper subgrade
Montana	Montana expansion index (swelling test)	Bituminous membranes	Conventional pavement sections

(Table continued on p. 168.)

TABLE 5.8. (*Continued*)

State	Design Criteria	Subgrade Treatment	Special Design Features
Oklahoma	Oklahoma soil index (classification)	Lime treatment, 4 to 6%	PCC sections—24 in. (60 cm) select borrow or lime-treated subgrade, 4 in. (10 cm) base, 9 in (23 cm) reinforced concrete Flexible sections—6 in. (15 cm) lime-modified subgrade, 9 in. (23 cm) base, 4.5 in. (11 cm) AC
South Dakota	AASHTO	Strict moisture-density control Lime treatment upper 6 in. (15 cm), 5–16%	Full depth asphalt 12 in. (30 cm) thick placed directly on treated subgrade
Texas	PVR Design curves based on subgrade shear strength	Lime treatment upper 6–12 in. (15–30 cm) Strict moisture density control Bituminous moisture barriers	Conventional pavement sections Subbase thickness design
Utah	AASHTO	Subexcavate 4 to 5 ft (1.2 to 1.5 m) and replace with select material Existing shale subgrade scarified and recompacted at 2% above optimum Bituminous membranes	Conventional pavement sections
Wyoming	AASHTO	Swell pressure tests	Subexcavate up to 5 ft (1.5 m) and recompact to near AASHTO 6-99 optimum Bituminous membranes Remove and replace if feasible Full depth asphalt

Summarized from Snethen et al. (1975).

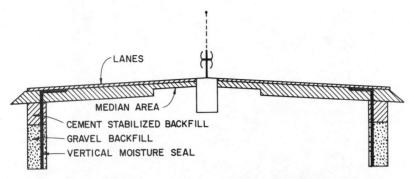

FIGURE 5.24. Deep vertical moisture barriers used on a San Antonio freeway.

in trenches at the edge of the pavement to act as vertical moisture barriers. The membrane was overalpped a distance of 2 ft (600 mm) into the asphalt paved shoulder, tacked with emulsion, and covered with asphalt concrete surfacing. Figure 5.25 shows the highway section before construction and a typical section along the same stretch of highway in 1985.

The second test section was similarly constructed along a 2-mi (3.2-km) section of Interstate 37 in 1980. The main lanes of this eight-lane divided highway were constructed on 6 in. (150 mm) of lime-stabilized subgrade, 8 in. (200 mm) of cement stabilized base, and 8 in. (20 mm) of concrete pavement. Both test sections were instrumented with moisture sensors. Profilometer and photologging test results as of December 1982 showed that in all cases the test sections provided a smoother riding surface than the control sections. Visual observations as of 1987 have indicated the successfulness of the vertical barriers in reducing roughness.

5.4.2.2.3 San Antonio Sand Underdrain Project (Krazynski, 1980). An underdrain, 8 ft (2.4 m) deep, was placed along one side of the roadway on U.S. Highway 90 west of San Antonio and backfilled with sand. The intent was to have the sand act as a moisture reservoir for the clay, thus facilitating a more stable moisture environment with seasonal fluctuations. Measurements taken since 1975 indicate that the movement along the underdrained edge is about one-third the movement along the unprotected edge. Movements measured in a nearby area located away from the sand underdrain test section indicate movements approximate five times as large.

5.4.2.2.4 Mississippi State Highway Department (Teng and Clisby, 1975; Teng et al., 1972, 1973; Krazynski, 1980). A test project was undertaken by the Mississippi State Highway Department to evaluate some construction methods that would eliminate or greatly reduce the heave of the subgrade after the completion of the roadway. In Phase I of the work, completed in 1970, a 0.85-mi (1.35-km) section was constructed consisting of two cut and two fill areas. The subgrade material was Yazoo clay, a highly expansive montmorillonitic clay. The typical soil properties were liquid limit = 98%, plasticity index = 65%, and montmorillonite

(a)

(b)

FIGURE 5.25. I-410 in San Antonio (a) before and (b) after installation of moisture barriers.

content = 65%. Mitigation techniques employed in the test section included flooding with the use of boreholes, lime treatment to produce capillary barriers to prevent desiccation, and a bituminous horizontal membrane.

During construction, the entire experimental section was undercut 3 ft (1 m) below finish and subgrade. In the section to be subjected to ponding, a grid of 6-in. (150-mm)-diameter boreholes was drilled to a depth of 20 ft (6 m) and filled with sand. The spacing of the boreholes was 5 ft (1.5 m) in both directions. A highly fissured system of cracks caused by desiccation was expected to transport the water between boreholes. The ponded area was flooded for 140 days.

Another portion of the experimental section was coated with an asphaltic membrane to seal off the surface from the intrusion of rain water. The section was instrumented with swell plates and access tubes for nuclear moisture probes. The entire area was then backfilled.

During the observation period after construction, the control section exhibited significantly more distortion than the treated sections. The program indicated that both ponding and bituminous seals were an effective method of reducing swell. Conclusions from the program can be summarized as follows:

- Flooding with the use of sand-filled boreholes was effective in Yazoo clay. The moisture apparently penetrated the soil well through holes spaced on 5 ft (1.5 m) centers, due to the system of fissures and cracks. The results indicated that while undisturbed clay in a cut section had a strong response to flooding, remolded clay in a fill section did not. That suggests that the presence of the fissures and cracks was important in providing water access throughout the soil mass. It is interesting to note that the nuclear probe access tubes installed prior to flooding were pinched off by the pressure of the swelling clays and consequently no accurate moisture measurements were obtained.

- The horizontal asphaltic seal near the surface was effective in reducing the swell. When the experiment was designed, it was originally anticipated that water would be introduced into the swelling soil from surface runoff and also by capillary action. Since no swelling was observed under the bituminous membrane, it was concluded that no moisture gain took place by capillary rise in the highly fractured Yazoo clays. The reason for this is not clear but it is evident that capillary rise was not effective in increasing the water content under the pavement above what existed prior to construction. This is probably a reflection of the fact that infiltration from precipitation exceeded evapotranspiration losses in the natural state. It may be concluded, therefore, that the principal mechanism by which moisture was introduced to the fractured Yazoo clays was by surface runoff.

5.4.2.2.5   Clifton–West Experimental Test Program: Colorado Department of Highways (Brakey, 1973). This project included 19 test sections along 5 mi (8 km) of I-70 in the Mancos Shale near Grand Junction. Ten test sections in bases of cuts were subexcavated by approximately 2 ft (0.6 m) and backfilled with various types and gradings of granular materials. Eight of these sections were instrumented with 20-ft (6-m)-deep aluminum moisture probe access tubes. Four sections had various treatments with hydrated lime, including drilled holes. One test section was prewetted, scarified to a depth of 3 ft (1 m) and recompacted. Two sections were covered with catalytically blown asphalt membranes, prior to placement of the base courses and pavement. There were several control sections. The results of several Colorado projects will be discussed together below. The principal findings of this project included the following:

- All subexcavated sections covered with granular backfill (and no membrane) showed some expansion. The sections that swelled the most were those covered by open-graded, granular material.
- Hydrated lime appeared to be of little benefit in this experiment.
- The two sections covered by ⅜-in. (10-mm)-thick, horizontal asphalt membranes outperformed all other preventative measures. The subgrades below the membranes remained close to the construction moisture content with little evidence of swell.

5.4.2.2.6   Ordway Experimental Project: Colorado Department of Highways (Brakey, 1973). Twenty test sections were placed over a 2.4-mi (3.8-km)-long stretch of State Highway 96 in southeastern Colorado. The purpose of the project was to evaluate different full-depth stabilized base materials. Four different materials were used: untreated aggregate, low stability hot sand–asphalt mix, high stability hot sand–asphalt mix, and asphalt concrete. Two inches (50 mm) of asphalt concrete wearing course was used on all sections. All control sections had 18-in. (450-mm)-deep untreated aggregate base courses. The asphalt base sections varied in thickness between 5.5 to 8.5 in. (140 to 215 mm). The general results will be discussed for all Colorado projects below. The principal results of this project included the following:

- All full-depth stabilized sections performed well, even several years after their design life.
- Moisture studies showed that the subgrades under the asphalt bases remained significantly drier than the control sections.

5.4.2.2.7   Elk Springs Project: Colorado Department of Highways (Brakey, 1973). This project was located in northwestern Colorado on U.S. Highway 40. Fourteen test sections along a 7-mi (11-km) stretch were constructed. Four sections had 7 in. (75 mm) of asphalt stabilized base, four had 3.5 in. (90 mm), and four had no base of any kind. Two sections had 12-in. (30-cm)-thick clay bases enveloped with catalytically blown asphalt membranes and covered with 2 in. (50 mm) of asphalt emulsion-treated sand base. The wearing course was 2.5 in. (63 mm) of asphalt concrete throughout. Catalytically blown asphalt membranes were placed in the ditches of all cut sections, reaching from the pavement to the bottom of the ditch and up the backslope to the level of the pavement surface. The principal findings of this project were as follows:

- The test sections constructed with a 3.5 in. (9 cm) or thicker asphalt stabilized base and membrane-lined ditches performed well.
- Sections constructed with no base and the membrane encapsulated test sections experienced structural and heaving failures.

5.4.2.2.8   Conclusions from Colorado Test Programs. From the Colorado highway test section programs the following conclusions can be summarized:

- Catalytically blown asphalt membranes approximately $3/16$ in. (0.5 cm) thick are economical and effectively stabilize the moisture in expansive shale subgrades.
- Full-depth asphalt bases may be effective in stabilizing moisture content in subgrade soils.
- Granular, untreated bases should be avoided as they are directly or indirectly responsible for higher moisture contents in the subgrades.
- Encapsulation, or envelopment of the pavement in an impermeable membrane may be effective if properly constructed, but sufficient support may not be economically gained when compared with full-depth asphalt.

Several other states besides those discussed above have also tried innovative subgrade treatments and design features in experimental sections. These states include South Dakota (McDonald and Potter, 1973), Wyoming (Brakey, 1973), Louisiana (Higgens, 1965), Oklahoma (Hartronft et al., 1969), and others. The success or failure of any technique depends upon soil conditions, climate, and construction quality control.

5.4.2.3 *Airfield Procedures*

The major differences between design criteria for highway and airport pavements are related to the magnitudes of loads and tire pressures, repetition of load applications, distribution of traffic, and the geometry of the pavement. In general, civilian airports within the United States are designed in accordance with the Federal Aviation Administration (FAA) standards, and military airfields in accordance with the U.S. Army Corps of Engineers' CBR analysis. The Asphalt Institute has also published a design manual (MS-11) and a computer program for full-depth asphalt airfield pavements intended for air carriers over 60,000 lb (27,000 kg) in gross weight. In addition, the Portland Cement Association (PCA) has developed a design manual and design charts for determining rigid pavement thicknesses for various aircraft (Packard, 1973). All of the design procedures include either specific or general recommendations for expansive subgrades. These recommendations pertain to grading operations, compaction and moisture control, and lime or other additive modifications of the subgrade.

One of the most extensive studies of airfield pavement design intended specifically for expansive soil subgrades was initiated by the FAA in 1975. The project was conducted as a joint effort with the U.S. Air Force Engineering and Services Laboratory (AFESC) at the University of New Mexico Engineering Research Institute. The results of the NMERI study are published in a series of reports, concluding with a design recommendation manual (McKeen, 1976; McKeen and Nielsen, 1978; McKeen, 1981).

The rational design procedure is presented in McKeen (1981). This procedure is the most comprehensive, rational approach to date for designing airfield pavements for expansive soils.

5.5 SUMMARY OF ENGINEERING PRACTICES FOR DESIGN ON EXPANSIVE SOILS

In the course of gathering data for preparation of this manuscript and as part of a research project sponsored by the National Science Foundation, personnel from several different consulting engineering firms, government agencies, and universities were interviewed. The information obtained from these interviews is summarized in Table 5.9 in terms of field and laboratory investigation procedures, design criteria, and the structural and soil treatment alternatives used in each area. It is important to note that Table 5.9 represents the interpretation and summarization of the interviews by the authors and is not intended to represent any endorsement of, or by, the people interviewed. The names of the companies, agencies, or universities are included in the table primarily to provide them recognition for their help.

TABLE 5.9. Summary of engineering practice in United States and Canada for investigation and design for expansive soils[a]

Location	Company or Agency Interviewed	Field and Laboratory Investigations				Design Structural Alternatives	Treatment Alternatives
		Reconnaissance	Preliminary	Detailed	Criteria		
Los Angeles, California	Moore and Taber Consulting Engineers and Geologists Slosson San Diego Soils (ICG)	Known geologic sources Visual inspections well-trained field personnel (engineers or technicians) considered essential	Atterberg limits Expansion index (EI) tests	CS tests	% swell from EI test PCA for pavements (no design criteria for flexible pavements)	Reinforced slabs on grade, monolithic and post tensioned types Narrow continuous footings placed 24 in. (60 cm) deep Pier and grade beams infrequent	Prewetting common Lime treatment has not had good success for buildings in area Lime treatment used extensively in highway design Remove and replace
Denver, Colorado	Chen and Associates F.M. Fox and Associates CTL/Thompson Geotek Consultants	Known geologic sources Expansive soils maps (Hart, 1974) Visual	Atterberg limits Dry density Some X-ray analysis % passing #200 SPT Atterberg	CS tests under surcharge pressures between 300 to 2000 psf (14 and 96 kPa),	% swell and swell pressure from CS tests Dry density Atterberg limits,	Drilled, straight shaft piers, grade beams with voids and floating floor slabs Narrow,	Some lime treatment (not popular in area) Moisture barriers in highway

Location	Firm						
	Woodward–Clyde Consultants	inspections Known geologic sources Field classification by engineer	limits in conjunction with CS tests	typically 1000 psf (48 kPa)—some samples air dried before swell CS tests, 1000 psf (48 kPa) surcharge—no reload to get swell pressure—no air drying prior to swell	correlated with CS results Pavement design by classification only CS and classification data reviewed by experienced engineer—use judgment in design Min. D. L. for piers 15000 psf (718 kPa)	continuous perimeter footings Isolated pads Best option is drilled piers with structural floor Recommended roughened piers and shear ring cutoffs below active zone	design Remove and replace Some lime treatment
Gainesville, Orlando, Florida	Schmertmann and Crapps Bromwell and Carrier, Inc. Ardaman and Assoc. Univ. of Florida	Known geologic sources Visual inspections	Atterberg limits w_n in comparison with PI	CS tests at overburden pressure	PTI criteria for slab design	Reinforced slabs-on-grade, post tensioned most common Belled piers and grade beams	Remove and replace
Minneapolis– St. Paul, Minnesota	Twin Cities Testing STS	Known geologic sources	Atterberg limits w_n in comparison	CS tests at overburden plus	% swell and swell pressure from	Narrow, deep spread footings	Piers and beams structural Do not

(*Table continues on p. 176.*)

TABLE 5.9. (Continued)

Location	Company or Agency Interviewed	Field and Laboratory Investigations			Criteria	Design Structural Alternatives	Treatment Alternatives
		Reconnaissance	Preliminary	Detailed			
	Consultants		with $w_n < w_p$, $w_l > 45\%$, high swell potential	surcharge, or 1000 psf (48 kPa) preswell loading	CS tests		recommend lime treatment for buildings, in general. Do not recommend prewetting
Fargo, North Dakota	Foss Assoc.		SPT Menard pressuremeter		Similar to Colorado design criteria		Remove and replace common in Fargo area
Oklahoma City, Norman, Oklahoma	Standard Testing and Engineering Co. Shepherd Engineering Testing Co.	Known geologic sources	Atterberg limits % passing #200 Natural moisture (w_n) in comparison with plastic limit to indicate trouble: $w_n < w_p$ indicates potential for heave	CS tests at overburden plus surcharge CV swell tests	General swell potential from CS and CV tests, qualitative expansion ratings and experience used to select design	Pier and beam, straight shaft or belled. Reinforced slabs on grade; waffle type or monolithic. Continuous footings	Remove and replace. Lime treatment for houses: slurry injection to 5–6.5 ft (1.5–2 m) depths, mixed in place in upper 9 in. (works best in "blocky" soils)

| Tulsa, Oklahoma | Stewart-White Associates Hemphill Corp. | Known geologic sources | PVC (used for pipe lines only) Atterberg Limits Linear shrinkage ratio (infrequently used) PVC (infrequently used) | CS tests at overburden plus foundation load surcharge pressure | Pier depth based on q_u of bedrock material PI > 28% % swell from CS tests Oklahoma Dept. of Transportation —Engr. Classific. of Geol. Mat'ls —for highways | Pier and beam Very little stiffened slab construction | Lime treatment common, amount based on PI reduction; do not recommend injection method Remedial underpinning, slab repair very common |
| Dallas–Ft. Worth– Arlington, Texas | Freese and Nichols Rone Engineers Structural Engineering | Known geologic sources Vegetation type | Atterberg limits Dry density % passing #200 TDHPT cone penetration and SPT Menard pressuremeter | Constant volume (CV) or controlled strain swell tests | PVR tests Pier uplift forces based on Atterberg limits Flexible pavement design based on PI | Reinforced slabs–on– grade Drilled, underreamed piers with floating floor slabs Drilled, straight | Remove and replace Lime treatment common Water pressure injection Moisture barriers |

(*Table continues on p. 178.*)

TABLE 5.9. (Continued)

Location	Company or Agency Interviewed	Field and Laboratory Investigations			Criteria	Design Structural Alternatives	Treatment Alternatives
		Reconnaissance	Preliminary	Detailed			
			PVR tests		Concrete pavement design based on CBR of stabilized subgrade material	shaft piers in swelling bedrock formations	
						Full depth asphalt pavements common	
	Corps of Engineers, Ft. Worth District	Known geologic sources	Consolidation tests according to procedures in TM 5-818-1 (Army, 9161) and EM 1110-2-1906 (Army, 1970)	Program HEAVE (Johnson 1979, 1982) to predict movement for foundation design	Pier and beam	Do not recommend lime injection for buildings in area	
		Atterberg limits Initial suction measurements in situ			Reinforced slab-on-grade		
			Soil suction tests (see Section 4.3, WES method)			Lime treatment in 15–30 cm (6–12 in.) of pavement subgrades	

Regina, Saskatoon, Saskatchewan	Clifton and Assoc. University of Saskatchewan	Known geologic sources Aerial photos Topography and landform identification	Atterberg limits National Research Council, Canada, Tech. manuals for identification (Hamilton, 1966): based on w_n, classif., and climatic factors	CV tests	Swell pressure from CV tests Volume change versus effective stress relationship	Reinforced slabs-on-ground Shallow belled piers with floating slabs most common, some deep straight piers and structural slabs	Lime stabilization for pavement subgrade treatment

CS = Consolidation-Swell Tests
CV = Constant Volume (Controlled Strain) Tests
SPT = Standard Penetration Test
EI = Expansion Index
PCA = Portland Cement Association
PTI = Post Tensioning Institute
PVC = Potential Volume Change
PVR = Potential Vertical Rise
TDHPT = Texas Dept. Highways and Public Transportation

6

TREATMENT OF EXPANSIVE SOILS

6.1 GENERAL CONSIDERATIONS AND GUIDELINES

Treatment procedures that are available for stabilizing expansive soils before and after construction of structures and highways include

- chemical additives
- prewetting
- soil replacement with compaction control
- moisture control
- surcharge loading
- thermal methods

Preliminary site investigations and evaluation of soil properties, as discussed in previous sections, must be undertaken before choosing an appropriate soil treatment. Both the climate above the ground surface and the microclimatology of the soil may affect the shrink–swell of the soil. Breakage of water drains, local watering practices, poor surface water drainage, local transpiration characteristics, and the location of such heat sources as fireplaces can affect soil shrinkage and swelling. Field and laboratory tests should be conducted to classify soil, identify properties, and develop a soil profile of the zone affecting the foundation. If expansive soils are present, the swell potential should be evaluated using one of the recommended procedures outlined earlier.

The successful application of soil stabilization procedures requires considerable experience and judgment regarding the soils on-site, consideration of limitations of the methods to be chosen, and correct implementation procedures. Treatment methods should be chosen after consideration of the following factors:

- Project size.
- Phasing of construction funding (i.e., the ponding treatment requires funds early in the project).
- Nature and use of the project (i.e., variations in strength requirements and maximum volume changes that can be tolerated for an airport runway, secondary highway, or residential house).
- Cost comparisons of alternative methods.

Currently used stabilization procedures will be discussed below in terms of effectiveness, economy, and ease of implementation. Limitations of each procedure also will be discussed. When appropriate, case studies of treatment methods applied to various construction projects will be included to aid the user in assessing his or her projects. Some methods of particular interest in highway construction will be referenced. A brief discussion of soil treatment methods was also presented in Chapter 5 with regard to case histories for highway pavements.

Guidelines should be followed for monitoring the field data and evaluating the effectiveness of treatment methods. Snethen (1979a) offered some sound suggestions for the construction of highways on expansive soils. These suggestions can be generalized as well to other types of construction. These guidelines are summarized in the following paragraphs.

Direct observation by qualified personnel can be as beneficial and more cost effective than systematic engineering measurements. Nevertheless, treatment alternatives should be evaluated using sound engineering judgment based on comparative data, not personal opinion. Minimum standards should be maintained for field monitoring data.

The measure of a treatment's success is the degree of control that it provides over both moisture content fluctuation and volume change in foundation or subgrade soils. The extent to which the treatment method is effective in controlling seasonal variations should be evaluated. Measurements between treated and control sections should always be compared.

Other factors are important for highway and/or structure performance:

- Subgrade strength. Some treatments, such as prewetting, will reduce the soil strength. Therefore the effect of strength loss must be evaluated to ensure that strength criteria are met.
- Surface deformation. Grid level surveys are useful if the time required to gather the data is not prohibitive. Statistical comparisons between treated and untreated sections are helpful in evaluating the treatment if a sufficiently large data base is available.

- Pavement performance. The treatment can be evaluated by comparing the serviceability indices or surface roughness for treated and untreated sections.

6.2   SITE PREPARATION

The preparation of the site can have as important an influence on the overall performance of the structure as the construction itself. Rough grading and construction of cut and fill areas may reveal soil characteristics not found previously in the site investigation program. It is very important that a qualified geotechnical engineer inspect the site after grading and/or excavation to verify or amend previous exploration results. Preferably this should be the engineer who did the initial site investigation. It is particularly important to check for soil irregularities such as thin lenses, and layers or pockets soil different from those that were indicated by the preliminary investigation.

Vegetation also can affect soil moisture and, in turn, affect the construction site. Tree and shrub roots can dry out the soil to depths as high as 15 ft (5m) or more (Wise and Hudson, 1971; De Bruijn, 1973). Even heavy grass vegetation along a concrete pavement can contribute to loss of soil moisture in the subgrade, which can cause shrinkage and loss of support (Felt, 1953). Removal of major root systems larger than ½ to 1 in. (12 to 25 mm) in diameter is recommended. As a general rule, plants over about 6 ft (2 m) in height should be removed.

6.3   REMOVAL AND REPLACEMENT

Removal of expansive soils and replacement with nonexpansive soils is one method to provide stable foundation material. In some cases, the expansive strata may be entirely removed. Generally the expansive layer extends to a depth too great to economically allow complete removal and replacement. It then must be determined what depth of excavation and fill will be necessary to prevent excessive heave.

Appropriate swell tests should be conducted to evaluate the expected potential heave. An attempt should be made to duplicate conditions in the field and use undisturbed sampling techniques. The depth to which nonexpansive backfill should be placed will be governed by the weight necessary to restrain the expected uplift pressures and the ability of the backfill to mitigate differential displacements. No definitive guidelines for this depth have been developed. Chen (1988) recommends a minimum of 3 to 4 ft (1 to 1.3 m).

One mechanism by which the removal and replacement method mitigates expansive potential is by control of the moisture content in the underlying clay layer. Most of the seasonal moisture content fluctuation takes place in the nonexpansive backfill. If the clay exhibits a moderate to low expansion potential the reduced volume of expansive clay in the upper, high moisture variation layer may be sufficient to prevent large movements at the surface. However, if there is a high potential for volume change in the underlying soil the reduced volume of expansive material may not adequately prevent surface heave or shrinkage. If water can infiltrate into the clay layer due to surface runoff, water line breakage, or other factors, excessive movements will most likely result.

Removal and replacement of expansive material have been successful in the repair of some hydraulic structures to stabilize uplift pressures. Repairs were made on the Friant-Kearn Canal (Holtz and Gibbs, 1956) and the Mohawk and Welton Canals by overexcavating the subgrade and replacing the expansive material with sand and lightly compacted gravel (Holtz, 1959). Since some of the swelling had already occurred, this method was an adequate remedial measure that provided not only extra weight to counteract future expansion but also allowed for differential movement to be taken up by the gravel layer.

However, caution should be exercised in using granular nonexpansive fill to replace expansive soils. Highly permeable fill will provide access for water and a reservoir for seepage into expansive subgrades or foundation soils. Impermeable, nonexpansive fill is much more satisfactory. If large thicknesses of granular material are a part of a pavement design, permanent, positive drainage and moisture barriers should be provided to minimize moisture ingress. As noted in Chapter 5 in the case histories of highway pavements, those test sections with granular subbases tended to exhibit larger movements.

It is important to recognize the effect of soil removal, replacement, and application of a surcharge on soil microclimate. The results of consolidometer tests on undisturbed samples of subgrades may not be accurate if the tests are performed under initial environmental conditions and changes in the soil system resulting from the use of the above treatment methods are not taken into account. For example, swelling characteristics may be altered during the soil removal due to a change in stress history or soil temperature.

Some advantages of treatment by removal and replacement include the following.

- Nonexpansive soils can be compacted at higher densities, yielding higher bearing capacities than can be produced by prewetting the expansive clay or compacting it at low densities.
- The cost of soil replacement can be more economical than other stabilization procedures since it does not require special construction equipment such as discs, harrows, mixers, or spreaders.
- Removal and replacement require less delay to construction than some other procedures such as prewetting.

Some disadvantages of removing and replacing the expansive soil are listed below.

- Nonexpansive, preferably impervious fill must be obtained. This can have a significant cost factor if the fill must be imported.
- The required thickness of the nonexpansive fill material may be too great to be practical.

• Granular fill may serve as a reservoir and provide a long-term source of water to foundation or subgrade soils.

6.4 REMOLDING AND COMPACTION

The swell potential of expansive soils can be reduced by decreasing the dry density. Figure 6.1 (Holtz, 1959) shows that compaction at low densities, and at water contents above the optimum water content, as determined by a Standard Proctor test, produces less expansion potential than compaction at high densities and low water contents. Better control of swell potential has been achieved by compacting the soil at water contents at or above optimum using a minimum density to reduce swell pressure (Chen, 1988; Dawson, 1959; Gizienski and Lee, 1965).

Design specifications should be clearly stated and implemented in the field. Laboratory tests should attempt to model conditions in the field. Moisture contents and densities should be within a specified range, and quality control tests should be performed often to ensure uniformity of the fill conditions. Selection of a water content slightly wet of optimum, based on compaction tests for the soil, will dictate a corresponding dry density that must be achieved. This can be translated to a minimum compactive effort produced by a certain number of passes of specified equipment needed to produce this dry density. It is difficult to overcompact the soil at water contents above optimum, but it is also difficult to compact stiff clay at

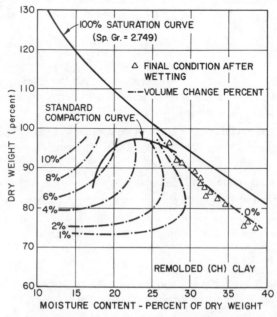

FIGURE 6.1. Percent expansion for various soil placement conditions under a load of 1 psi (Holtz, 1959).

water contents 4 to 5% above optimum moisture. If moisture contents are too high, clays can become so sticky and plastic that equipment cannot handle the soil properly and the soil can lose significant bearing capacity.

Ripping or scarifying the soil requires less effort than subexcavation if the depth of influence is less than 2 ft. The method may be adequate to use for secondary highway construction if larger deformations can be tolerated. In some cases ripping or scarifying the subgrade may be recommended for primary highways if the swell potential is low (Snethen, 1979a). Compaction criteria should always be implemented.

The specified soil conditions should be checked prior to construction to ensure that the soil has not dried out. If there is expected to be a long time delay between soil placement and structure placement, then means to prevent water content changes may be necessary. Ponding of the surface is not recommended because this can cause slaking of the surface and produce difficult working conditions.

Some advantages of remolding and recompacting include the following.

- The clay on-site can be used as fill thereby eliminating costs of importing fill.
- If the compaction is performed properly, the impermeable fill will minimize migration of water into underlying soils.
- It is economically feasible to scarify, pulverize, and recompact expansive soils.

Some disadvantages of remolding and recompaction control include the following.

- The lower bearing capacity of the low-density compaction may not be adequate. However, for light structures, such as single family houses, the bearing capacity of clays compacted at low densities is generally adequate.
- Some soils have such a high potential for volume change that compaction control does not reduce swell potential significantly and replacement may be necessary.
- Compacting at specified densities and water contents may necessitate frequent testing to maintain quality control, which may increase the cost of the project.

Experience has shown that good moisture-density control is essential for good engineering results. Used in conjunction with lime stabilization or soil replacement it provides excellent control of expansive clays for many projects. Subexcavation and remolding are most effective for soils having a natural water content dry of optimum and a high natural dry density. Highly cracked soils would benefit by the creation of a more uniform, impermeable soil surface.

6.5 SURCHARGE LOADING

Swell can be prevented if expansive clays can be loaded with a surcharge large enough to counteract the expected swell pressures. This is generally applicable only for soils with low to moderate swelling pressures. Field and laboratory tests should be conducted to determine the swell characteristics of the soil. The conditions in the field should be approximated during testing. As swell pressures increase, the use of a surcharge becomes less efficient because of the nonlinear nature of the pressure–swell relationship.

Judgment must be used to effectively apply a surcharge to control potential swell of the foundation soils. The surcharge method is most effective when swell pressures are low and some heaving can be tolerated in the construction project, such as in a secondary highway system. For large projects involving high foundation pressures and in which anticipated swell pressures are low to moderate, this method may also be effective.

Many soils exhibit swell pressures too high to be controlled by normal surcharge loads alone. Swell pressures up to about 600 lb/ft^2 (25 kPa) can be controlled by 4 ft (1.3 m) of fill and a concrete foundation. However, some soils may have swell pressures as high as 8000 lb/ft^2 (400 kPa).

6.6 PREWETTING

Prewetting or ponding is based on the theory that increasing the moisture content in the expansive foundation soils will cause heave to occur prior to construction and thereby eliminate problems afterward. It is assumed that if the high moisture content is maintained, there will be no appreciable increase in soil volume to damage the structure. This procedure may have serious drawbacks that limit its application. Expansive soils typically exhibit low hydraulic conductivity and the time required for adequate wetting can be up to several years. Furthermore, after the water has been applied for long periods of time serious loss of soil strength can result causing reductions in bearing capacity and slope stability. Another major drawback to the use of this procedure is that after a prolonged period of surface ponding the wetting front of the infiltrating water will have advanced to only to a depth much less than that of the active zone. Redistribution of water throughout the active zone can continue after construction due to the high water content in the zone above the wetting front. The continued migration of water into lower layers can result in continued heave after construction.

However, the experience of geotechnical engineers in the southern California area has indicated that in some cases the problem soils have sufficiently high hydraulic conductivity that the method has been used with some success. Also the case history in Mississippi discussed in Chapter 5 indicated that prewetting was effective in fissured clay. (It is also important to note that it was not effective in compacted clay that was not fissured.)

A commonly used procedure for prewetting is to construct dikes or small earth berms to impound the water in the flooded area, as shown in Figure 6.2. Trenches also have been constructed below the foundation and then flooded (Dawson, 1959). In the test case in Mississippi (Chapter 5) holes were drilled to provide access for water to depths.

The results of most ponding projects show that ingress of water usually occurs very slowly in dense clay soils. Highly fissured clays respond more favorably to

(a)

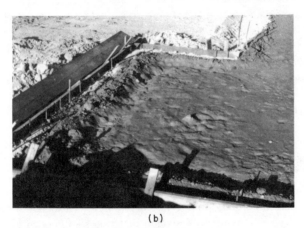

(b)

FIGURE 6.2. Prewetting treatment of foundation soil: (a) overview of site and (b) close up of edge berm (courtesy of Moore and Taber Consultants, California).

ponding because the fissure system allows water to reach a greater volume of the soil mass. Prewetting has greater chances of success during the hot, dry season when soils are in a desiccated state.

Very long periods of time may be necessary to sufficiently wet dense clay soils if the ponding method is used alone. Periods of ponding time up to 1 to 4 months may be sufficient to increase moisture contents to an adequate level in fairly permeable soils. In some cases, however, time periods up to a year or more may be necessary to cause effective penetration of the wetting front.

Prewetting to moisture contents 2 to 3% above the plastic limit has provided acceptable results for slab-on-ground foundations (Poor, 1979). There are some indications that the use of a surfactant may increase the percolation rate, but more research needs to be done in this area before conclusions can be drawn.

A complete site investigation and full regime of soil tests should be conducted prior to implementing the ponding treatment. Since this method may be limited to use in soils with low to moderate expansion potential a good assessment of the expansion potential should be made. The anticipated expansion is related to initial and final moisture contents of the soil. Therefore, targeted moisture contents associated with maximum swell should be known before prewetting the foundation material. The effect of the ponding procedure should be analyzed by performing periodic field tests to determine moisture contents at depths of interest in the soil. The depth of the clay layer that is expected to contribute to heave, or the active zone, will determine the necessary depth to test. Instrumentation of the site may be useful to monitor the amount or rate of swell and changes in moisture content during the prewetting period. Elevation monuments installed at different depths and nuclear probes have been used successfully (Teng et al., 1972; Porter, 1977; Nelson and Edgar, 1978; Goode, 1982; Hamberg, 1985).

Installation of vertical sand drains can reduce the time needed for water penetration and heave by reducing the length of the flow path. This technique was used by Blight and deWet (1965). The wells should be drilled on a grid pattern. The main factors to be considered are the coefficient of permeability, the active depth of the clay layer, the area affecting the construction, the radius of the sand drains, and the number of drains installed. Other soil properties such as clay mineralogy and fabric may influence the rate of swell. The in situ coefficient of permeability may vary from that calculated from laboratory tests, especially with desiccated, highly fissured soil. Therefore, field tests or past experience may be helpful in developing a grid pattern.

The prewetting method can take very long periods of time to accomplish. Nevertheless, in some cases it has been successful if sufficient time exists to allow the soil to swell. Teng et al. (1972) described a procedure used to reduce the swell potential of a highly fissured Yazoo clay formation in Mississippi. Sand drains 20 ft (6 m) deep were used and ponding lasted 140 days. However, it also must be noted that an adjacent remolded fill section that was also flooded did not exhibit the same positive results as the in situ soil. The lack of success in the fill area can be attributed primarily to the compacted soil having a much lower hydraulic conductivity than the in situ soil due to lack of fissures and a different soil fabric due to the compaction.

In another project, Blight and deWet (1965) were able to accomplish approximately 90% of the estimated maximum heave in a clay that was prewetted for 2 months using sand drains. Felt (1953) was able to increase moisture contents in desiccated, fissured clay soil within a ponding period of 1 month. However, continued flooding up to 6 months produced continued swelling. Steinberg (1981) used the ponding technique on a section of highway in Texas. He concluded that the method was successful in causing most of the expected heave prior to construction after approximately 1 month of ponding. McDowell (1965) ponded a section of Waco, Texas highway without using sand drains and was able to achieve water penetration up to 4 ft deep in 24 days.

Van der Merwe et al. (1980) successfully stabilized expansive fissured soils for road construction in South Africa using a prewetting procedure. A sand blanket was placed over the site to be used as a reservoir for irrigation water over a period of 50 days. The degree of saturation was increased down to a depth of 40 in. and at least half of the expected heave was developed. Ponding was used as a remedial measure to stabilize the subgrade soils of the Friant-Kearn Canal (Holtz and Gibbs, 1956). The ponding technique has been coupled with lime stabilization of several inches of the surface soil to produce a firm working base for equipment operation. The lime increases the strength of the top layer of soil and helps to prevent evaporation in the subgrade.

The reader is cautioned, however, against being unduly influenced by the apparent success of this method as indicated by the frequency of successful applications being documented in the literature. There have been many failures resulting from loss of soil strength and the creation of unsuitable working conditions due to excessive moisture in the upper soil layers. Unfortunately, only the successes are reported in the literature and the failures tend to be suppressed.

6.7 CHEMICAL ADMIXTURES

6.7.1 Lime Stabilization

Lime stabilization has been used successfully on many projects to minimize swelling and improve soil plasticity and workability. Many state highway departments have researched lime stabilization and frequently use this treatment method. Generally, from 3 to 8% by weight hydrated lime is added to the top several inches of the soil. It is also used as a follow-up treatment over ponded areas to add strength to the surface, provide a working table for equipment, and prevent evaporation by capillary rise from the treated layer below (McDowell, 1965; Teng et al., 1973).

The chemical theory involved in the lime reaction is complex (Thompson, 1966, 1968). The primary main reactions include cation exchange, flocculation–agglomeration, lime carbonation, and pozzolanic reaction. All fine-grained soils can be modified to some degree to exhibit less plasticity and improved workability using lime treatment. The strength characteristics of a lime-treated soil depend primarily on soil type, lime type, lime percentage, and curing conditions (time–temperature).

6.7.1.1 Soil Factors
Factors influencing a soil's lime reactivity are listed below (Thompson, 1966).

- A soil pH greater than about 7 indicates good reactivity.
- Organic carbon greatly retards lime–soil reactions.
- Poorly drained soils tend to have higher lime reactivity than well-drained soils.
- Calcareous soils have good reactivity.
- Sulfates and some iron compounds inhibit the lime reaction (Currin et al., 1976).
- The presence of gypsum in the soil or ammonium fertilizers may increase the amount of lime required (O'Neill and Poormoayed, 1980).

Pozzolanic reactions can occur in some soils and depend on soil properties such as clay content, clay mineralogy, soil pH, organic content, and drainage characteristics. Other reactions that produce stabilization of soil include the production of strong cementing agents from lime, water, and aluminous or siliceous substances. The solubility of silica in clay soils is increased in the high pH environment created by the addition of lime (McKeen, 1976). The lime also supplies a divalent calcium cation that can form calcium silicates and calcium aluminum hydrates, which can increase soil strength. As mentioned previously, organics, sulfates, and some iron compounds inhibit the reactions (Moore and Jones, 1971).

Some caution should be used when attempting to stabilize clays that contain gypsum. In the presence of water and lower temperatures, hydration of gypsum and crystallization of expansive minerals can cause heave.

Soils that are highly weathered and are better drained are less reactive than poorly drained soils in the temperate regions. An indicator of the degree of weathering is the Ca/Mg ratio. Weathering tends to leach calcium ions from the system reducing the Ca/Mg ratio. Other indicators of a more highly weathered condition are a low pH and low base saturation. A low base saturation shows that there are fewer free hydrogen ions. Better drained soils may have iron existing in the colloidal state, which may disrupt the pozzolanic reaction (Currin et al., 1976). In tropical or subtropical regions, a better index of lime reactivity is provided by the silica–alumina ratio (Hardy, 1971).

Quicklime or hydrated lime can improve the engineering properties of heavy clay soils or granular soils with silt–clay fractions. Clay–gravel material has been successfully stabilized for use as pavement bases. Lime treatment is not recommended for sandy soils with no fine fraction and is not very effective with silt–loam soils. Fly ash or other pozzolanic material can be added to most granular soils to improve the gradation and reactivity of the soil (Winterkorn, 1975).

6.7.1.2 Testing Procedures

6.7.1.2.1 Eads and Grim pH Test. Eads and Grim (1966) have developed a quick test to determine if a soil is lime reactive and how much lime, in percent by weight, is necessary to achieve a desired volume change reduction. The procedure is shown in Table 6.1 and can be completed in less than 2 hr. The results of the test can be plotted on a graph showing percent lime versus pH. The lowest percent lime to produce a pH of 12.4 is termed the "lime modification optimum," or LMO. It is recommended that Atterberg limits be determined in conjunction with pH tests, at least within 2% on either side of the LMO.

A drawback of this method is that a pH of 12.4 does not ensure lime–soil reactivity (Currin et al., 1976). Other factors must be considered to assess the effectiveness of lime in reducing the expansion potential. One such indicator is the Plasticity Index (PI). If the PI is not reduced by 50% at the LMO, then lime is probably not a practical modification method (Snethen, 1979b). For soils having a low potential for volume change (PI of 35 or less) the reduction in PI should be 15 at the LMO. Another indicator is the soil strength. The soil is considered reactive if there is an increase in unconfined compressive strength of at least 50 psi (350 kPa) (Thompson, 1966).

TABLE 6.1. Suggested pH test procedure for soil–lime mixtures

Materials

   1. Lime and soil

Apparatus

   1. pH meter (the pH meter must be equipped with an electrode having a pH range of 14)

   2. 150-ml (or larger) plastic bottles with screw-top lids

   3. 50-ml plastic beakers

   4. Distilled water (CO_2 free)

   5. Balance

   6. Oven

   7. Moisture cans

Procedure

   1. Standardize the pH meter with a buffer solution having a pH of 12.45

   2. Weigh to the nearest 0.01 g representative samples of air-dried soil passing the No. 40 sieve and equal to 20.0 g of oven-dried soil

   3. Pour the soil samples into 150-ml plastic bottles with screw-top lids

   4. Add varying percentages of lime, weighed to the nearest 0.01 g, to the soils. (Lime percentages of 0, 1, 2, 3, 4, 5, 6, and 8, based on the dry soil weight, may be used)

   5. Thoroughly mix soil and dry lime

   6. Add 100 ml of CO_2-free distilled water to the soil–lime mixtures

   7. Shake the soil–lime mixture and water for a minimum of 30 sec or until there is no evidence of dry material on the bottom of the bottle

   8. Shake the bottles for 30 sec every 10 min.

   9. After 1 hr, transfer parts of the slurry to a plastic beaker and measure the pH

  10. Record the pH for each of the soil–lime mixtures. The lowest percent of lime giving a pH of 12.40 is the percent required to stabilize the soil. If the pH does not reach 12.40, the minimum lime content giving the highest pH should be used

From Eads and Grimm (1966).

6.7.1.2.2 Soil Stabilization Index System (SSIS). The SSIS was developed at Texas A&M University and is a collection of procedures developed for use in pavement maintenance and construction. A detailed account of the use of the SSIS is in Currin et al. (1976). Procedures for cement and asphalt stabilization are also included in addition to those for lime stabilization. Figure 6.3 shows the procedure for determining the most appropriate method to be used for stabilization. It is based on the PI of the soil. In brief, the SSIS procedure is as follows.

- Determine natural soil properties.
- Estimate the optimum lime content (LMO).
- Determine whether the soil is lime reactive.
- Determine the durability of the soil using the estimated percentage of lime. Durability tests include rapid cure at several temperatures, freeze–thaw tests,

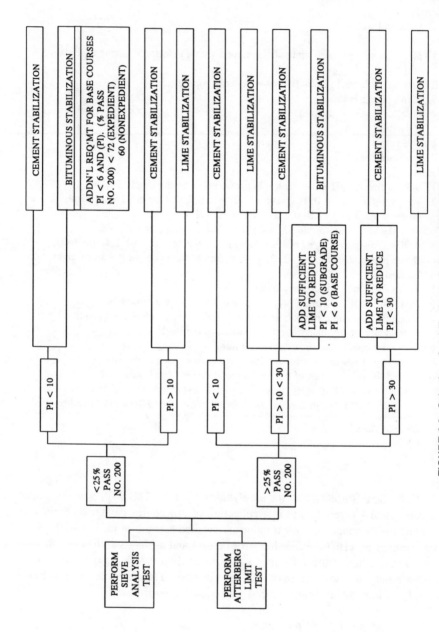

FIGURE 6.3. Selection of stabilization technique—SSIS Method.

tests showing strength development over time, the effect of sulfates and organics, and vacuum saturation strength tests to determine the immersed strength.

- Verify the above tests and finalize the design.

6.7.1.3 Mixture Design Procedure (Thompson, 1968)

Thompson (1968) developed design procedure to be used for highway construction on Illinois soils. In the course of developing the procedures he came to several conclusions about lime stabilization.

- All medium and fine-textured Illinois soils were improved to some degree by lime treatment.
- Reactive soils stabilized with high quality lime produce strong, durable subbase and base material for highway construction.
- Nonreactive soils, producing little strength increase, still have some improvement of characteristics such as workability and CBR. Economically lime treatment may be justified in some cases.

Thompson's (1968) design procedures are shown in Figure 6.4.

6.7.1.4 Type and Amount of Lime

Table 6.2 lists several types of lime used as additives (Winterkorn, 1975). High calcium hydrated lime has a somewhat higher degree of effectiveness but lime type does not greatly influence the plasticity, workability, or expansion potential (Thompson, 1968).

The type and amount of lime used can be important if the strength of the soil is also a consideration. Thompson (1968) ranked three types of hydrated lime in order of their effectiveness on increasing the soil's strength. The limes he considered were, in order of effectiveness, dolomitic lime, by-product high calcium lime, and high calcium lime. The U.S. Army Corps of Engineers has experienced better results using quicklime (CaO) rather than hydrated lime, but due to possible environmental problems associated with quicklime, government authorities discourage its use (Skinner, 1973).

For high quality lime the percentage used is usually between 3 and 7%. Thompson (1968) found that treatments of 5 and 7% lime produced higher strength gains than 3% percent lime, and the strength gain at 73°F increased up to 56 days during curing. Gokhale and Swaminathan (1973) cured black cotton soil from India with 10, 12, and 14% lime and 1 and 2% of sodium silicate at each lime level. The lower percentage of lime (10%) achieved the highest gains in strength, and samples with 14% lime exhibited the lowest gains. The results also indicated that the addition of 1% sodium silicate accelerated strength development above levels achieved with either 2% or no addition of sodium silicate.

These results indicate that there exists an optimum lime content at which maximum soil enhancement is achieved. It may be expected that the optimum content will differ whether one is concerned with strength or expansion potential. Approximate hydrated lime and quicklime contents for clayey gravels, silty clays, and clays are

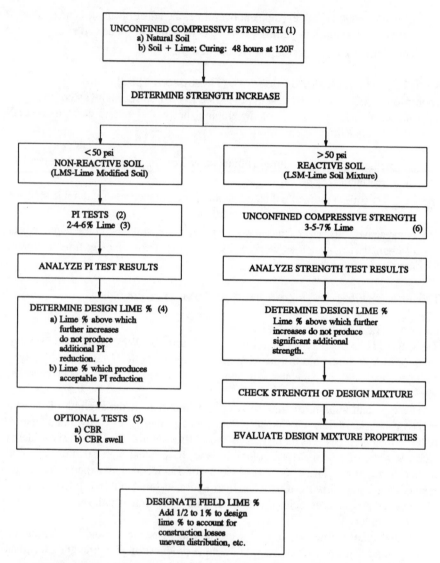

FIGURE 6.4. Flow diagram for proposed mix design process (Thompson, 1969b).

given in Table 6.3 (Currin et al., 1976). The percentages given were developed at Texas A&M for use in the SSIS methodology.

6.7.1.5 Curing Conditions

Increased temperature and time improve the gain in strength for lime-stabilized soils, and may be expected to have some influence on expansion potential as well.

It is advisable to time the construction to obtain maximum benefit of summer temperatures before the onset of cold weather. If the soil temperature is less than

TABLE 6.2. Lime materials used in construction

Type	Formula
Calcia (high-calcium quicklime)	CaO
Hydrated high-calcium lime	$Ca(OH)_2$
Dolomitic lime	$CaO + MgO$
Normal hydrated or monohydrated dolomitic lime	$Ca(OH)_2 + MgO$
Pressure hydrated or dehydrated dolomitic lime	$Ca(OH)_2 + Mg(OH)_2$

60° to 70°F and is not expected to increase for 1 month, then the chemical reactions will be deterred and the benefits will be minimal (Currin et al., 1976). Alternate methods of stabilization should be used if cold environmental conditions are expected. The curing period should extend for at least 10 to 14 days before heavy vehicles are allowed on the lime-stabilized soil.

6.7.1.6 Application Methods

6.7.1.6.1 Mixed in Place and Recompacted. One method of lime application is to mix it mechanically with either a disc harrow or a small ripper. With this method it difficult to mix deeper than about 12 in. (300 mm). At the Dallas–Fort Worth Airport Project, an 18-in. (450-mm) layer was stabilized in two successive lifts. The top 9 in. (225 mm) was moved aside so the bottom lift could be stabilized. A second method was also used to mix to a depth of 18 in. (450 mm). A root plow was used on the back of a D-8 bulldozer to cut the entire thickness at once (Kelly, 1973).

Lime can be applied dry, which is fastest, or in a lime slurry. Dry spreading, however, can create problems such as bulking of the soil and air pollution. Also lime must have an excess of water for a proper reaction to occur. It is important that specifications be established and procedures be followed to ensure that they are met. Quality control of lime content, pulverization, mixing, and compaction

TABLE 6.3. Approximate lime contents for various soil types

Soil Type	Approximate Treatment Percent by Soil Weight	
	Hydrated Lime	Quicklime
Clayey gravels (GC, GM-GC)(A-2-6, A-2-7)	2–4	2–3
Silty Clays (CL) (A-6, A-7-6)	5–10	3–8
Clays (CH) (A-6, A-7-6)	3–8	3–6

must be maintained (Mitchell and Raad, 1973). The soil should be prepared by initially breaking the layer to be treated into clods of a specified size. After the lime is mixed into the desired depth of the subgrade, it should not be compacted until the size of the clods of the stabilized material is small enough to allow sufficient mixing and adequate compaction. The top layer can then be sealed by rolling and compacting to prevent fluffing due to carbonation reactions or precipitation. After 3 or 4 days the material can be mixed for the last time with discs on a pulvimixer and then compacted with a sheepsfoot or pneumatic roller to specified moisture and density values. If the surface is sealed with an asphalt emulsion, the moisture content will stabilize in the lime-treated material over a period of time and pozzolanic reactions can take place.

The depth of application for mix-in-place lime is limited. Therefore, more success can be expected when the active zone is shallow, and if only the top layer would contribute significantly to the total heave. Lime-treated subgrades of as thin as 4 to 6 in. have been successful in protecting underlying layers that have been subexcavated and replaced using moisture-density control (McDonald and Potter, 1973).

Normally, however, a significantly thicker layer must be treated and if the heave is deep-seated, a thin stabilized layer may be ineffective. In some cases thicknesses up to 3 ft (1 m) have been treated.

Deep plowing of lime has been successfully used in Oklahoma (Thompson, 1969a; Hartronft et al., 1969). Large tractors rated at 200 horsepower or larger equipped with heavy ripper blades were able to mix the lime to a depth of 2 ft (600 mm). A lower percentage of lime may be required for deep lime stabilization than for 6-in. (150-mm) stabilized layers.

The mix-in-place method of lime stabilization has been reported to be effective in minimizing swell and increasing the strength of prewetted soil if applied to lime-reactive soil using proper specifications. The method is widely used in highway construction, but has also been used effectively and economically for foundation pretreatment. Precautions should be taken to prevent the access of surface or ground-water to the lime stabilized layer.

6.7.1.6.2 *Drill Hole Lime.*

The drill-hole technique consists of introducing quicklime or hydrated lime in a slurry form into holes drilled for the purpose. This method has been used as a remedial measure in highway stabilization (Snethen 1979a; Thompson and Robnett, 1976). Holes 6 to 12 in. (150 to 300 mm) in diameter are drilled through the pavement to depths of 30 to 50 in. (750 to 1250 mm), depending on the nature of the problem zone. Smaller diameter holes provide less surface area to contact the slurry. Hole spacings may be 4 to 5 ft (1.2 to 1.5 m) centers. Water is introduced in the lime slurry or with dry lime to increase the mobility of the lime.

Results of the drill-hole technique are erratic, and the authors do not encourage its use. One factor that limits the effectiveness of the method is the inability to uniformly distribute the lime in the soil mass (Thompson and Robnett, 1976). Also the diffusion process is very slow unless the soil has an extensive network of fissures. Snethen (1979b) indicated that water migration and stress relief were the major mechanisms affecting vertical deformation instead of the lime. Brakey (1973) reported negative results using the drill-hole technique on Mancos shale near Clifton, Colorado. On the other hand, Gerhardt (1973) concluded that the drill-hole technique in Colorado highway construction was effective in reducing swell potential at any depth. Differences in various factors such as soil type and texture or quality control during construction undoubtedly account for disparity in results using the drill-hole method.

A variation of the drill-hole technique was used to construct a 4-story slab-on-grade structure in Texas (Haynes and Mason, 1965). Trenches, 6 in. (150 mm) wide and 3 ft. (1 m) deep, were installed on a grid pattern at approximately 10 by 30 ft. (3 by 9 m) centers. The trenches were filled with 1 ft. (300 mm) of commercial hydrated lime, then with gravel, water, and a surfactant. It took 2 to 3 weeks to increase the moisture content in the upper 10 ft. (3 m) of clay. This procedure limited swell in the subsurface clay and was significantly more economical than construction of a structural floor slab.

In general, a lower degree of confidence will exist for success of the drill-hole technique. In low permeability materials both the moisture migration and lime penetration will be minimal. The same limitations as apply for prewetting will also apply to this method. For soils that are highly fissured, however, the introduction of a lime slurry may effect a degree of moisture stabilization that may be effective. Before using this method it is recommended that evidence of success in local practice be sought.

6.7.1.6.3 Pressure-Injected Lime. The pressure-injected lime (PIL) or lime slurry pressure injection (LSPI) technique was developed in an attempt to produce greater lime slurry penetration in the drill-hole method. In this method lime slurry is pumped through hollow injection rods at pressures at intervals of about 12 in. (300 mm). Slurry is injected until either the soil will not take additional slurry, or until injection begins to fracture or distort the surface. The nature of the soil will affect the quality of slurry that can be injected. An average take is about 10 gal/ft. Field experience has shown that a slurry composition of 2½ to 3 lb of lime per gallon of water is satisfactory (Thompson and Robnett, 1976). Center spacings are commonly 3 to 5 ft. (1 to 1.5 m). Current equipment is capable of injecting to depths of 10 ft. (3 m).

Even with the high injection pressures field tests have indicated that penetration of injected lime slurry occurs only along planes of weaknesses or fissures and the slurry will not penetrate soil pores (Higgins, 1965; Ingles and Neil, 1970; Lundy and Greenfield, 1968; Wright, 1973). The procedure is most effective at times of maximum desiccation of the soil mass. If the soils are highly fissured, and are lime reactive, the injection of the lime can "seal off" the zones of clay between the fissures and produce a stabilizing effect on the moisture content between fissures.

From the literature and from general discussion with engineers the authors became aware of many reports of both successes and failures. Local practice and success rates should be taken into account before considering these methods.

6.7.2 Cement Stabilization

The hydration of Portland cement is a complex pozzolanic reaction that produces a variety of different compounds and gels. The result of mixing cement with clay soil is similar to that of lime. It reduces the liquid limit, the plasticity index, and the potential for volume change. It increases the shrinkage limit and shear strength (Chen, 1988).

However, Portland cement is not as effective as lime in stabilizing highly plastic clays. Some clay soils have such a high affinity for water that the cement may not hydrate sufficiently to produce the complete pozzolanic reaction. It is usually advantageous to use cement when soils are not lime reactive (Mitchell and Raad, 1973).

For clays in which Portland cement stabilization is effective, mixing procedures similar to those used for lime stabilization can be used. One difference in technique is that the time between cement addition and final mixing should be shorter than that used for lime treatment. Portland cement has a shorter hydration and setting time. Because of the strength increase that can be generated by the use of cement, the soil–cement mixture can increase the pavement and slab strength significantly. A 2 to 6% cement content can produce a soil that acts as a semirigid slab. This will aid in distributing expansion more uniformly throughout the slab. However, cement stabilized materials may be prone to cracking and should be evaluated for this effect prior to use. AASHTO specifications exist for cement requirements for soils in the AASHTO Soil Groups.

Since the mixing methods for lime and cement are nearly the same, so are the processing costs. Overall treatment costs may be similar for lime and cement (Portland Cement Association, 1970).

6.7.3 Salt Treatment

The most common salts used in soil stabilization are sodium chloride and calcium chloride. The effect of sodium chloride on soil properties is variable. It generally has a greater effect in soils having a high liquid limit. Depending on the soil type, sodium chloride may increase the shrinkage limit and shear strength. For soils that react with sodium chloride, there may also be some beneficial control of frost heave.

Calcium chloride will also stabilize moisture content changes in soils, thereby reducing the potential for volume change. It has been used to control frost heave in soils since the late 1920s. About 1% calcium chloride by weight of dry soil is needed to stabilize most soils.

Disadvantages of using calcium chloride are that it is easily leached from the soil, and the relative humidity must be at least 30% before it can be used.

There is insufficient evidence that salts other than sodium chloride and calcium chloride have adequate soil stabilization capabilities to be economically justifiable. Due to leaching, salt treatment generally must be repeated every 3 years or so. This temporary nature of the treatment may make this method not economically viable (Gromko, 1974).

6.7.4 Fly Ash

Fly ash has also been added to soils treated with lime to increase the pozzolanic reaction and improve the gradation of granular soils. The pozzolanic activity of silty soils has been improved by using a lime-fly ash ratio of 1:2 (Woods et al., 1960). There are, however, a wide variety of types of fly ash having different mechanical and chemical properties. Therefore, for a specific application a comprehensive testing program would be needed to determine the design criteria necessary for fly ash stabilization.

6.7.5 Organic Compounds

A variety of organic compounds have been tried as soil stabilizers. No organic compound, however, has proven to be as effective or economical as lime. There are proprietary compounds on the market that may have specific applications. Experimentation in the field with these products is necessary however before they can be recommended.

The success of using organic compounds is limited. Various organic compounds have been used for waterproofing, retardation of water adsorption, or hardening of the soil with resins.

Chen (1988) describes in detail field tests of proprietary compounds called Fluid 705, 706, and 707. Even though laboratory tests indicated some of the fluids were effective in controlling swelling and plasticity, the results of injecting the fluids into holes under pressure were disappointing.

Experience with pressure injection of organic compounds have indicated that many organic compounds react quickly and irreversibly and are not water soluble. Their use during prewetting procedures is doubtful (Mitchell and Raad, 1973). The diffusion rates in expansive soils are very low making spraying or injection ineffective. Therefore, direct mixing is the only appropriate method of application.

6.8 MOISTURE CONTROL BY HORIZONTAL AND VERTICAL BARRIERS

Soil expansion problems are primarily the result of fluctuations in water content. Nonuniform heave will result from either nonuniform water content changes, nonuniform soil conditions, or both. If fluctuations in water content over time can be minimized and if the water content in the subsoils can be made uniform a major part of the problem can be mitigated.

The placement of a structure or pavement on the ground surface will change the evapotranspiration from the surface. Changes in land use, such as irrigation of landscaping, will change the potential for infiltration. These changes will, in turn, change the water content and its distribution in the subsoils. If the changes in water content can be made to occur slowly and if the water content distribution can be made uniform, differential heave can be minimized.

Moisture barriers offer a viable solution in this regard. The basic principle on which moisture barriers act is to move edge effects away from the foundation or pavement and minimize seasonal fluctuations of water content directly below the structure. Also, the time during which moisture changes occur is longer because the barrier increases the path length for water migration under the structure. This allows for water content to be more uniformly distributed due to capillary action in the subsoil. Thus the heave will occur more slowly and in a more uniform fashion.

Moisture barriers have been used both as a preconstruction technique and as a remedial measure. Methods of construction are important in the success of moisture barriers. Care should be taken to seal joints, seams, rips, or holes in the barrier. Deep-rooted plants should be planted well beyond the perimeter of the moisture barrier.

6.8.1 Horizontal Moisture Barriers

Horizontal moisture barriers, installed around a building or on highway shoulders, limit the migration of moisture into the covered area. Guidelines for successful applications of horizontal barriers have been presented by Woodward-Clyde-Sherard (1968) and Hammitt and Ahlvin (1973). Horizontal barriers can be categorized as membranes, rigid barriers, or flexible barriers. The width of the barrier should be at least as wide as the edge distance discussed previously with regard to slabs on grade. Some factor of safety added on to the width would be appropriate.

6.8.1.1 Membranes

Impermeable sheets in varying grades and thicknesses have been used as horizontal membranes. Polyethylene, ranging from 4 to 20 mil in thickness, polyvinyl chloride (PVC), polypropylene, high-density polypropylene, and other types of nonwoven fabrics have been used with varying degrees of success. Information on longevity and long-term chemical stability should be obtained from the manufacturer prior to installation.

Membranes with thicknesses below 10 mil require special care to avoid puncture during placement (van der Merwe et al., 1980). Care must be taken to prepare the surface of the site. All sharp, projecting materials must be removed, as well as vegetation and organic material. The surface should be even and uniformly compacted at a specified density. A final rolling with a steel drum roller will aid in eliminating some sharp projecting particles.

Most organic membranes will degrade if exposed to sunlight. Therefore, barriers should be covered to protect them against ultraviolet radiation and damage by physical contact. Chemical compatibility of the membrane with the soil should be ensured. Care should be taken when dumping and spreading soil cover on the membrane to avoid damage from equipment or the dumped material.

The edges of the membrane should be secured so that leakage does not occur next to the foundation or at the membrane overlaps. Waterproof glue or mastic is recommended to attach the membrane to the foundation.

When membranes are used around the perimeter of a building they should be placed deep enough to prevent damage from root growth. Large plants should be eliminated or moved well away from the barrier to a distance at least 1 to 1½ times the height of the plant (Johnson, 1979).

There may be a tendency for moisture to accumulate at the outer edge of the membrane. Moisture migration into the foundation soils can be reduced by providing a subdrain at the membrane's edge, which can discharge water. The drains should be constructed with ample slope to avoid any chance of them backing up water. Surface water coming off the roof of the structure should be directed well away from the edge of the horizontal barrier. Figure 6.5 shows a typical building foundation moisture barrier.

From 1947 to 1950 a technique was utilized in Texas to prevent volume change in highway embankments constructed of high-volume change clays (Van London, 1953). The problem soil was encapsulated in asphalt membranes sprayed to an approximate thickness of ³⁄₁₆ in. (5 mm). Even though the asphalt application did not meet the standards established by the Bureau of Reclamation for treating canals and reservoirs, the encapsulated soil remained stable for more than 10 years after construction (Benson, 1959).

A similar technique, membrane-encapsulated soil layers (MESL), utilizing prefabricated membranes has been applied at the U.S. Army Engineer Waterways Experiment Station to control volume change (Hammitt and Ahlvin, 1973). The swell potential of the soil may remain high using this method, but will not be activated because water is prevented from migrating into the encapsulated soil layer. The configuration of the membranes is shown in Figure 6.6. For very deep layers of soil with a high swell potential, this method may not be economically feasible. However, thin to moderate thicknesses of problem soils may be stabilized using this technique.

The encapsulation technique was also used on highway sections in Colorado (Gerhardt, 1973). The subgrade moisture content was controlled. However, not enough additional soil strength was gained compared to soil under an asphalt base to justify the cost of placing the membrane.

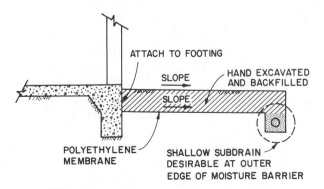

FIGURE 6.5. Typical detail of a horizontal membrane.

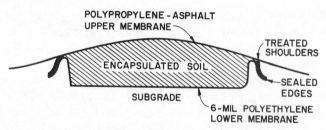

FIGURE 6.6. Typical cross section of a membrane encapsulated soil layer for airfields (Hamitt and Ahlvin, 1973).

The successful use of encapsulation requires careful quality control. Membranes must remain intact during and after placement, and joints must be properly sealed. If soil strength is an important factor, moisture-density criteria should be carefully controlled in the encapsulated soil.

6.8.1.2 Asphalt

Asphalt has been commonly used in highway construction as a membrane to prevent the intrusion of surface water into the subgrade. Catalytically blown asphalt can be an effective waterproofing material. The Asphalt Institute recommends that about 1.3 gal/yd^2 (5.9 liters/m^2) of asphalt cement should be used for a membrane. Prefabricated asphalt sheets less than ½ in. (12 mm) thick also have been used (Chen, 1988).

Asphalt membranes can be used in conjunction with other materials. When applied under a layer of nonexpansive fill, the likelihood of surface waters penetrating into underlying expansive soils is minimized. This method can be used under slab-on-grade projects and in swimming pool construction to decrease water infiltration and damage from heave.

When asphalt membranes are sprayed, the surface should be smooth and free of debris so that punctures are prevented. Structures such as sign posts and guardrails should be sealed so that leaks next to the membrane do not occur.

Asphalt membranes, particularly catalytically blown, emulsified, and asphalt-rubber, or asphalt bases are routinely used in highway construction and can be economical and effective in maintaining the moisture content at placement levels (Gerhardt, 1973; Brakey, 1973; Morris, 1973; Shiflett, 1974; Snethen, 1979a). Road tar and bitumen have been tried as flexible membranes. Van der Merwe et al. (1980) used a grade 50/55 road tar sprayed at the rates of 0.66 and 0.44 gal/yd^2 (3 and 2 liters/m^2). The lower application rate was as successful as the higher rate for an impermeable membrane. Bitumen was also tested, but it was difficult to spray evenly because the sand layer it coated could not be made sufficiently smooth.

Full depth plant-mixed bituminous bases have been used in Wyoming highway construction (Diller, 1973). These bases were placed directly on the subgrade or on membranes and are an economical alternative. Steps must be taken to prevent moisture from entering the subgrade at the pavement edges.

Various combinations of sprayed membranes and asphalt pavements over subgrades and ditches have been used (Snethen, 1979). Asphalt membranes should be applied over the entire subgrade, down the verge slopes and up the back slope to a vertical distance equal to about 1½ ft (400 mm) above the ditch invert or about 1 ft (250 mm) above the finished pavement grade (Snethen, 1979). Figure 6.7 shows examples of a continuous sprayed asphalt membrane and full depth asphalt pavement. If economically feasible, median strips of four-lane highways should also be included in the asphalt membrane treatment if they are less than two lanes in width (Snethen, 1979).

Full-depth asphalt pavements usually perform better than concrete when placed over highly expansive subgrade soils (Snethen, 1979a). The asphalt pavement is flexible, thereby allowing more distortion before severe damage occurs. Because of a lower degree of cracking of the asphalt it is more effective as a waterproof membrane to reduce surface water infiltration. Also the remedial repair of asphalt pavement is less complex and easier to accomplish.

Paved shoulders also serve as horizontal moisture barriers. AASHTO specifications require a 10-ft (3-m) shoulder on outer lanes and a 4-ft (1.2-m) shoulder near median strips. A preferable width for the inner shoulder is 6 to 8 ft (2 to 2.5 m) (Snethen, 1979).

6.8.1.3 Rigid Barriers

Concrete aprons or sidewalks are commonly used as horizontal moisture barriers around building foundations. Concrete, coupled with a flexible membrane or asphalt, has been used successfully as a remedial measure to establish constant uniform

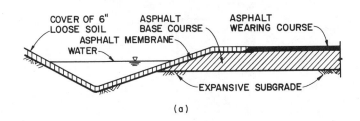

(a)

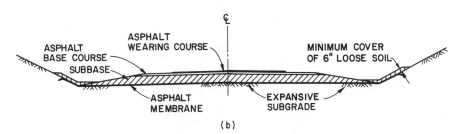

(b)

FIGURE 6.7. (a) Full-depth asphalt pavement and (b) typical sprayed asphalt membrane applications to minimize subgrade moisture variations from surface infiltration (Snethen, 1979).

moisture contents within foundation soils (Mohan and Rao, 1965; Lee and Kocherhans, 1973; Najder and Werno, 1973).

The design and function of joints and seals are important when using a rigid horizontal barrier. Heaving can occur at the edge of the apron, and if it tilts toward the structure, surface water will be directed to the edge of the foundation and into foundation soils. A flexible sealed joint should be provided between the structure and the barrier. Concrete sidewalks should be reinforced and attached to the structure with dowels to prevent the barrier from becoming detached. Sidewalks and aprons should have adequate slope away from the foundation so that even with some distortion flow will occur outward from the structure. Horizontal barriers will be more efficient if surface drainage is provided so that ponding is prevented.

6.8.2   Vertical Moisture Barriers

Vertical moisture barriers function in much the same way as horizontal barriers in terms of slowing the rate of heave and causing the water content distribution to be more uniform below the structure. However, vertical barriers are more effective than horizontal barriers in retarding lateral moisture migration. Consequently, edge effects are minimized.

Goode (1982) constructed field test plots to evaluate vertical moisture barriers. Four test plots were constructed, two of which had vertical barriers and two of which did not. A general cross section of the vertical barrier is shown in Figure 6.8.

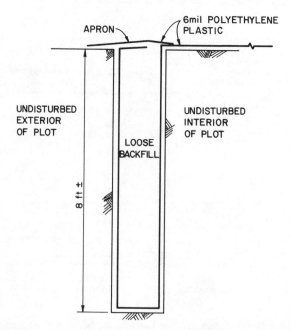

FIGURE 6.8. Moisture barriers used in experiments by Goode (1982).

The test plots were built on the Colorado State University campus at a location where the subsoils consisted of highly expansive Pierre clayshale. Previous investigations at the site indicated that the active zone extended to a depth of between 12 and 17 ft (4 and 5 m) (Nelson and Edgar, 1978).

Aluminum tubes were installed within the test plot areas and outside of the test plots to depths of about 20 ft. These tubes provided access for a neutron moisture probe. Survey markers to monitor heave were placed on a grid over each test plot. The test plots were covered with 6 in. of sand and gravel over a polyethylene liner to simulate a floor slab. The results are described in detail in Goode (1982).

In general, the results indicated that for the first year or two the heave under the slabs with vertical barriers was much less than that under the slabs with no barriers. After 4 to 5 years the maximum heave in both the slabs with and without vertical barriers was nearly the same. However, the differential heave observed under the slabs without the vertical barriers was much greater than that observed in the slabs with vertical barriers (Hamberg, 1985). The vertical barriers were effective in slowing the rate of heave and in decreasing the amount of differential heave even though the eventual total heave was not affected.

Vertical barriers should be installed at least as deep as the zone that is most affected by seasonal moisture change. It is generally not practical to place them to the full depth of the active zone, but a depth of one-half to two-thirds of the active zone is recommended. Barriers less than 2 to 3 ft (0.6 to 1 m) deep in general will not be sufficiently effective to warrant their placement (Snethen, 1979a). If the barriers cannot be installed at the time of construction they generally must be placed 2 to 3 ft (0.6 to 1 m) away from the structure to allow space for the excavating equipment to maneuver. In that case horizontal barriers must be installed to connect the vertical barrier to the outer edge of the structure.

The use of sand backfill in the vertical barrier trenches may provide a reservoir that eventually can leak through the membrane (Newland, 1965). The backfill material should be as impermeable as is possible. The effectiveness of the barrier will be improved if adequate surface drainage away from the structure is provided.

The types of materials that can be used for vertical barriers include impermeable membranes such as polyethylene, concrete, and impervious semihardening slurries. Membranes should be durable enough to resist puncturing or tearing during placement. Trenching machines have been observed to be more efficient for excavating than backhoes.

Poor (1979) described the use of three types of vertical moisture barriers. These included a capillary barrier composed of coarse limestone, a lean concrete barrier, and a barrier consisting of asphalt and ground-up rubber tires. These vertical barriers were observed to be successful in decreasing differential movements.

Vertical barriers can also be applied in highway construction. Without barriers water contents will increase toward the center and decrease at the edges causing pavement distortions if the subgrade is a high volume change soil. Figure 6.9 shows a typical highway application.

The Texas highway department has experimented using vertical waterproof membranes (Steinberg, 1981). The membranes were installed as a remedial measure,

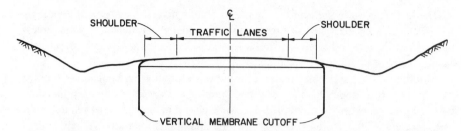

NOTE: THE DEPTH OF THE MEMBRANE CUTOFF SHOULD EXTEND TO THE DEPTH OF ACTIVE ZONE; HOWEVER, IF THE SOIL CONDITIONS AND THE PRACTICALITY AND ECONOMICS OF THE INSTALLATION ARE NOT FAVORABLE THEN THE DEPTH SHOULD BE REDUCED TO FACILITATE INSTALLATION.

FIGURE 6.9. Example of a vertical membrane cutoff in highway construction (Snethen, 1979).

and results indicated that less differential heave was experienced within the protected areas. Observation of a test section of interstate highway in San Antonio, Texas has shown this treatment to be very effective even 8 or 9 years after installation of the vertical membranes. This has been discussed in Chapter 5.

6.9   ELECTROCHEMICAL SOIL TREATMENT

The method of electroosmosis and base exchange of clay materials is referred to as electrochemical soil treatment. Dewatering and hardening of the soil are two results of electroosmosis.

The purpose of electrochemical hardening is to alter the physicochemical properties of the swelling clay by introducing a high concentration of preferred exchangeable cations into the clay. The technique can reduce swell pressure and the percentage of swell (O'Bannon et al., 1976).

The technique requires electrodes to be placed in the soil mass. An electric potential between the anode and cathode facilitates the movement of a solution into the soil to act as a stabilizing agent. Injection wells are often drilled to allow the chemical solution greater surface area for access into the soil. Laboratory tests should be conducted to determine the most effective anode material, and the type and amount of electrolyte required for the particular soil. Variables, such as electrode spacing, current, and voltage gradients, should be established by experimentation.

Electrochemical hardening has been implemented with relative ease by unskilled and semiskilled personnel on sections of a heavy-flow Arizona highway experiencing swelling problems (O'Bannon et al., 1976).

Electrochemical soil treatment is generally economical only for very localized areas, such as highways where right of way is limited or beneath overpasses where maximum clearance is necessary (Snethen, 1979b).

6.10 HEAT TREATMENT

Little work has been done in the United States to study or apply thermal treatment to expansive soil. Heating clays to approximately 200°C can significantly reduce the potential for volume change (Aylmore et al., 1969; Post and Padua, 1969). However, practical, economical methods have not yet been developed (Mitchell and Raad, 1973).

6.11 SUMMARY OF SOIL TREATMENT METHODS

The expansive soil treatment alternatives presented here may be employed either singly or in combination, to control heave. However, it must be recognized that in any case, site-specific conditions could make one or more methods ineffective. The choice of the technique to be used should be assessed with respect to

- Economic factors.
- Relative expected control of volume changes by implementing different treatment alternatives.
- Site-specific conditions such as potential for volume change, moisture variations, degree of fissuring, permeability, etc.
- Nature of the project.
- Necessary strength of the foundation soils.
- Time frame available for treatment.

Geotechnical site investigations and testing programs are important in making a suitable selection of a treatment method. It is important that test conditions should duplicate field conditions. Some factors of special interest are

- Potential for volume change.
- Depth of active zone.
- Degree of fracturing.
- Heterogeneity or uniformity of soils on-site.
- Lime reactivity of the soil.
- Presence of undesirable chemical compounds.
- Moisture variation within the soil mass.
- Soil permeability.
- Strength of the soil needed for the project.

Table 6.4 presents a summary of the most salient points to consider when applying each treatment alternative. Value judgments concerning the preference of one alternative over another are omitted unless it is believed that general agreement exists on the specific issue by the geotechnical engineering community. The selection process remains a matter of applying judgment to comparisons of pros and cons for the selected alternatives with special reference to the factors mentioned previously.

TABLE 6.4. Salient points to consider for the successful application of expansive soil treatment alternatives

Method	Salient Points
Removal and replacement	Nonexpansive, impermeable fill must be available and economical
	Nonexpansive soils can be compacted at higher densities than expansive clay, producing high bearing capacities
	If granular fill is used, special precautions must be taken to control drainage away from the fill, so water does not collect in the pervious material
	Replacement can provide safe slab-on-grade construction
	Expansive material may be subexcavated to a reasonable depth, then protected by a vertical and/or a horizontal membrane. Sprayed asphalt membranes are effectively used in highway construction.
Remolding and compaction	Beneficial for soils having low potential for expansion, high dry density, low natural water content, and in a fractured condition
	Soils having a high swell potential may by treated with hydrated lime, thoroughly broken up, and compacted—if they are lime reactive
	If lime is not used, the bearing capacity of the remolded soil is usually lower since the soil is generally compacted wet of optimum at a moderate density
	Quality control is essential
	If the active zone is deep, drainage control is especially important
	The specified moisture-density conditions should be maintained until construction begins and checked prior to construction
Surcharge loading	If swell pressures are low and some deformations can be tolerated, a surcharge load may be effective
	A program of soil testing is necessary to determine the depth of the active zone and the maximum swell pressures to be counteracted
	Drainage control is important when using a surcharge. Moisture migration can be both vertical and horizontal
Prewetting	Time periods up to as long as a year or more may be necessary to increase

TABLE 6.4. (*Continued*)

Method	Salient Points
Prewetting (*continued*)	moisture contents in the active zone
	Vertical sand drains drilled in a grid pattern can decrease the wetting time
	Highly fissured, desiccated soils respond more favorably to prewetting
	Moisture contents should be increased to at least 2–3% above the plastic limit
	Surfactants may increase the percolation rate
	The time needed to produce the expected swelling may be significantly longer than the time to increase moisture contents
	It is almost impossible to adequately prewet dense unfissured clays
	Excess water left in the upper soil can cause swelling in deeper layers at a later date
	Economics of prewetting can compare favorably to other methods, but funds must be available at an early date in the project
	Lime treatment of the surface soil following prewetting can provide a working table for equipment and increase soil strength
	Without lime treatment soil strength can be significantly reduced, and the wet surface may make equipment operation difficult
	The surface should be protected against evaporation and surface slaking
	Quality control improves performance
Lime treatment	Sustained temperatures over 70°F for a minimum of 10 to 14 days is necessary for the soil to gain strength. Higher temperatures over a longer time produce higher strength gains
	Organics, sulfates, and some iron compounds retard the pozzolanic reaction of lime
	Gypsum and ammonium fertilizers can increase the soil's lime requirements
	Calcareous and alkaline soils have good reactivity
	Poorly drained soils have higher reactivities than well-drained soils

(*Table continues on p. 210.*)

TABLE 6.4. (*Continued*)

Method	Salient Points
Lime treatment (*continued*)	Usually 2–10% lime stabilizes reactive soil
	Soil should be tested for lime reactivity, and percentage of lime needed
	The mixing depth is usually limited to 12–18 in., but large tractors with ripper blades have successfully allowed in place mixing of 2 ft of soil
	Lime can be applied dry or in a slurry, but excess water must be present
	Some delay between application and final mixing improves workability and compaction
	Quality control is especially important during pulverization, mixing, and compaction
	Lime-treated soils should be protected from surface and groundwater. The lime can be leached out and the soil can lose strength if saturated
	Dispersion of the lime from drill holes is generally ineffective unless the soil has an extensive network of fissures
	Stress relief from drill holes may be a factor in reducing heave
	Smaller diameter drill holes provide less surface area to contact the slurry
	Penetration of pressure-injected lime is limited by the slow diffusion rate of lime, the amount of fracturing in the soil, and the small pore size of clay
	Pressure injection of lime may be useful to treat layers deeper than possible with the mixed-in-placed technique
Cement treatment	Portland cement (4–6%) reduces the potential for volume change. Results are similar to lime, but shrinkage may be less with cement
	Method of application is similar to mix-in-place lime treatment, but there is significantly less time delay between application and final placement
	Portland cement may not be as effective as lime in treating highly plastic clays
	Portland cement may be more effective in treating soils that are not lime reactive
	Higher strength gains may result using cement

TABLE 6.4. *(Continued)*

Method	Salient Points
Cement treatment *(continued)*	Cement stabilized material may be prone to cracking and should be evaluated prior to use
Salt treatment	There is no evidence that use of salts other than NaCl or $CaCl_2$ is economically justifiable
	Salts may be leached easily. Lack of permanence of treatment may make salt treatment uneconomical
	The relative humidity must be at least 30% before $CaCl_2$ can be used
	Calcium and sodium chloride can reduce frost heave by lowering the freezing point of water
	$CaCl_2$ may be useful to stabilize soils having a high sulfur content
Fly ash	Fly ash can increase the pozzolanic reaction of silty soils
	The gradation of granular soils can be improved
Organic compounds	Spraying and injection are not very effective because of the low rate of diffusion in expansive soil
	Many compounds are not water soluble and react quickly and irreversibly
	Organic compounds do not appear to be more effective than lime. None is as economical and effective as lime
Horizontal barriers	Barrier should extend far enough from the roadway or foundation to prevent horizontal moisture movement into the foundation soils
	Extreme care should be taken to securely attach barrier to foundation, seal the joints, and slope the barrier down and away from the structure
	Barrier material must be durable and nondegradable
	Seams and joints attaching the membrane to a structure should be carefully secured and made waterproof
	Shrubbery and large plants should be planted well away from the barrier
	Adequate slope should be provided to direct surface drainage away from the edges of the membranes

(Table continues on p. 212.)

TABLE 6.4. (*Continued*)

Method	Salient Points
Asphalt	When used in highway construction a continuous membrane should be placed over subgrade and ditches
	Remedial repair may be less complex than concrete pavement
	Strength of pavement is improved over untreated granular base
	Can be effective when used in slab-on-grade construction
Rigid barrier	Concrete sidewalks should be reinforced
	A flexible joint should connect sidewalk and foundation
	Barriers should be regularly inspected to check for cracks and leaks
Vertical barrier	Placement should extend as deep as possible, but equipment limitations often restrict the depth. A minimum of half of the active zone should be used
	Backfill material in the trench should be impervious
	Types of barriers that have provided control of moisture content are capillary barrier (coarse limestone), lean concrete, asphalt and ground-up tires, polyethylene, and semihardening slurries
	A trenching machine is more effective than a backhoe for digging the trench
Membrane encapsulated soil layers	Joints must be carefully sealed
	Barrier material must be durable to withstand placement
	Placement of the first layer of soil over the bottom barrier must be controlled to prevent barrier damage

7

REMEDIAL MEASURES

7.1 OVERVIEW

The design of foundations and pavements on expansive soils always involves a certain degree of risk of damage. Usually a lower cost of the design alternative will be associated with a higher degree of risk. The same principle applies to the design of remedial measures employed to correct damage that has taken place.

Obviously, the maximum cost of a remedial measure is what would be required to remove the damaged structure and/or pavement and reconstruct a new one. Even this procedure requires considerable engineering design. The foundation soils will have caused movement but it will be necessary to assess the current state of expansion and to predict future expansion potential that still may exist.

In the authors' research in preparation for this book, engineers were interviewed at locations from Canada to Texas, from the midwestern United States to the west coast, and in several countries such as Israel and South Africa. In almost all locations, there existed some case in which the cost of remedial repairs would exceed the cost of removing the structure and rebuilding it. Not all cases are that severe, however, and a variety of innovative techniques have been developed for cost-effective remedial measures.

Prior to undertaking remedial measures a number of important issues must be discussed among the owner, the engineer, and other parties that may be financially involved. These questions include the following:

- Should remedial measures be undertaken at this time? If damage is not severe and continued future movement is anticipated, it may be better to wait until the rate of movement has slowed.

- Should remedial measures be undertaken at all? If damage is not severe and remedial measures offer no significant improvement it may be appropriate to make cosmetic repairs and continue maintenance.
- Who is financially responsible for the repair?
- What criteria should be used to select the remedial measure and scope of repairs to be employed. These should include considerations of cosmetic and structural benefits as well as actual costs.
- What is the cause and extent of the damage?
- What remedial measures are applicable?
- What residual risk will exist after the repairs have been completed? It should be expected that future movement will continue and the cost/risk relationship of the selected remedial measures must be considered.

The economic responsibility to implement remedial measures may fall on the owner, contractor, insurance agent, or realtor. The party responsible for the repair generally makes the decisions regarding the choice of a remedial measure. However, that party may have only a limited period of liability, whereas the owner's interest is usually for an indefinite long-term period. It is important that the party or parties who will benefit most from the repairs should be in a position to at least influence the selection of the remedial measures to be undertaken. The role of the engineer is to define the cause of the damage, make recommendations for corrective action, and possibly approve or inspect completed remedial work.

The selection and application of technically effective and cost-effective remedial measures are influenced by the differing interest groups. An agreement must be reached to determine how much future risk, both actual and economic, is acceptable as compared to immediate costs and present values. An important point of contention is the definition of what constitutes a failure, or whether remedial measures are warranted and necessary. Since each interest group has differing objectives it is often necessary to reach a compromise to avoid future controversy. The ultimate liability of each group should be made clear.

The process of applying remedial measures will be greatly improved if all parties can agree on performance standards and assignment of responsibilities for the assumption of residual risks. Realistically this rarely occurs.

Long-term guarantees on remedial repairs are virtually impossible to realistically provide. It may be difficult to adequately evaluate the economic soundness of a particular remedial measure, but applying only cosmetic repairs (leveling, patching, redecoration) without appropriate analysis of causes and damage is often an unrewarding investment.

In choosing a remedial measure each case should be evaluated using the pertinent questions listed at the beginning of this section. It is important that the advice of a professional engineer be sought to outline possible alternatives, initial costs, residual risks, and benefits. The success of the remedial measures employed will depend on the knowledge and skill of the engineer and contractor and the degree of quality control that is exercised.

Alternative remedial measures discussed in this chapter will give the reader an indication of some methods that are available. Some recommendations are made regarding the repair of particular types of damages, but there exist many possibilities to applying remedial measures either singly or in unique combinations. Frequently, local contractors have developed innovative techniques that can be very successful in their area. Each case should be evaluated by a competent engineer to provide the owner with the necessary information on which to base a decision.

7.2 REMEDIAL MEASURES FOR BUILDINGS

7.2.1 Investigation of Structure and Foundation Soil

The first step in the investigation of structural damage should be a detailed on-site inspection by a professional engineer with experience relating to expansive soil. All history pertaining to the structure and the soil is valuable, but is sometimes difficult to obtain. When making the site inspection it is highly recommended that the engineer use a questionnaire or data-gathering form. One such form used by an engineering firm in Texas is shown in Figure 7.1. The form being used should be developed and customized for the local area and type of structure being investigated. The primary advantages of using a standard form of this type is that the data are gathered in a consistent manner, detailed information is not overlooked, and the data are assembled in an orderly and organized fashion. In addition to the information gathered on the form, elevation readings should be recorded around the foundation and slabs to aid in evaluation of the problem.

It is recommended that investigators develop their own form specifically for the area in which they practice. The form shown in Figure 7.1 is intended for use in a particular area where basements are typically not used. Observations for remedial recommendations in areas where basements are common are somewhat different. The following information should also be determined in areas where pier and grade beam foundations and basements are common (Chen, 1988).

- Type of foundation.
- Design criteria.
- Water table condition.
- Type of foundation soils.
- Moisture content of foundation soils.
- Swelling potential of foundation soils.
- Dead load pressure exerted on footings or piers.
- Size of footings or piers.
- Length of piers.
- Pier reinforcement, expansion joint, underslab gravels, and other details.
- Subdrainage system.
- Condition of slab and underslab gravel layer.

Sample Residential Investigation Checklist (Courtesy John McCabe, Structural Engineering, Arlington, Texas)

I.   Client or Owner

Name: Date:

Address: Phone: Location of Property:

II.   Legal Description of Property

Lot: Block: Subdivision:

III.   History of Structure

Age of structure: Date occupied:

Square Footage:

Any Additions: (Locate on sketch)

Water leaks: (Locate on sketch)

Cracks in Foundation: (Locate on sketch)

History of Repair: Who:

Sprinkler System: When Installed:

When Did Problems First Occur?

Describe Major Problems:

Builder: H.O.W.   #

IV.   Outside Observations (Locate all info on layout)

Construction Type: Pier and Beam: Slab-on-Grade

   Measured Beam Depth: INT. ____in.   EXT. ____in.   DIA. ____in.

Trees Near the Structure:

General Topography of Site: Hill: Flat: Grading Away From Foundation?

Trapped Water Situation Exists?

Do Dryout Areas Exist?   Does Structure Have Gutters?

Two-story Areas: Quadrant: N.W. ____   N.E. ____   S.W. ____   S.E. ____

Signs of Soil Movements on Local Streets:

   Rolling: Cracks: Numerous Repairs:

   New: Old: Holes:

Fill Conditions at Time of Construction (show on drawing)

Surface Soil (check nearby trenches, holes, etc.)   Clay: Sand: Rock: Other:

Materials Used on Exterior:

Fireplace or Masonry?   Structure Have Control Joints?

Does Brickwork Have Weep Holes?   Retaining Walls?

V.   Roof

Framing: Trusses: Joist: Tile:

Size of Rafters: Spacing: Span:

Spacing of Purlin supports: Length:

Ridge Support Lengths: Member:

Clips?   Diagonal Bracing?   Collar Ties?   Roof Problems?

VI.   What Are the Principal Causes of Movement?

VII.   Repair and Recommendations

VIII.   Interior Observations (Indicate only major cracks or problems) Quadrants

	N.W.	N.E.
	S.W.	S.E.

Areas	Quad.		Areas	Quad.	
Entry			Bedrooms		
Hall			Baths		
Kitchen			Utility		
Garage			Dining		
Living			Patios		
Den			Other		

IX.   Special Conditions or Comments

X.   Sketch Indicate:

Approximate distances, dimensions   Topography

North Arrow   Dry Out Areas

Fill Areas   Walls

Control Joints   Slab and Masonry Cracks

Trap Water

FIGURE 7.1. Sample residential investigation checklist.

- Condition of piers.
- Void space under grade beam.

The site investigation should include as much information about the foundation soils and water table as possible. The resources available for the soil investigation will obviously depend on the severity of the problem and the magnitude of the remedial program. In any case there are minimum data requirements to determine the cause and extent of the expansion. Because it is often difficult to determine if the structural distress is caused by settlement or heave, it is essential to determine the soil profile and the expansion potential of the foundation soils. Also, information such as sources of water and depth of natural water table is important.

In many cases, a significant amount of this type of information can be obtained from visual inspection and experience with soil conditions of the area. In many other cases, especially if the damage is great and remedial measures will be extensive, or if litigation is involved, a detailed soil investigation including sampling, testing, and in situ measurements will be required.

The presence of the structure or pavement usually contributes to changes in the water content condition and state of stress in the soil. Therefore, comparison of soil conditions and water contents adjacent to or below the structure or pavement with those at some location more remote and less affected by the structure will provide valuable information on the likely cause of expansion and the extent to which it may continue.

Also the investigation must determine the type of foundation system, evaluate the quality of the construction, and determine the degree of distress that the foundation elements have suffered. This will generally require that test pits be excavated next to the grade beam and interior supports. If drilled piers exist, one or more should be excavated to observe tension cracks, voids at the bottom, and quality of concrete, and to check the depth of the pier. Figure 7.2 shows tension cracks caused by uplift on a pier. Figure 7.3 shows a pier that was not constructed to sufficient depth. Figure 7.4 shows mushrooming at the top of a pier resulting from inadequate formwork during construction.

Although the excavations and removal of slabs may be disruptive they are frequently necessary in the course of stabilizing the foundation. They may or may not seriously impact the habitability of the structure depending on their extent and location and the tolerance of the inhabitants. It is necessary, however, to stress that such investigations are essential. If inadequate site investigations lead to wrong diagnoses and improper remedial measures, the resources saved actually represent a waste.

During soil testing, environmental conditions must be considered. The in situ moisture content and dry density can indicate soil conditions, but conducting consolidation-swell tests on soils that have already reached a maximum swell in the field will not reveal the initial soil condition and may be misleading. Comparison of test results on samples taken under or near the structure with those for samples taken from a nearby uncovered area may give an indication of expansion that has taken place. Alternatively, the foundation soil can be air dried and rewetted. Comparison

FIGURE 7.2. Horizontal tension cracks in a 24-in. (60-cm)-diameter pier which had 0.5% steel. These breaks occurred at 20 to 22 ft (6 to 6.7 m) below the ground surface due to uplift pressures in the surrounding expansive soil (courtesy U.S. Army Corps of Engineers, Fort Worth District).

FIGURE 7.3. The pier on the right was not placed sufficiently deep. A new pier had to be installed after excavating and jacking up the foundation (courtesy Fu Hua Chen, Chen and Associates, Denver, Colorado).

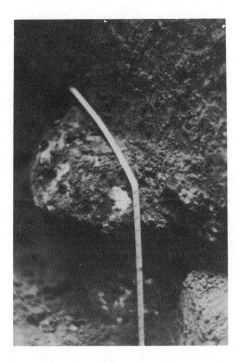

FIGURE 7.4. This pier was improperly placed. Excess concrete was left at the top, forming a "mushroom." Swelling soils under the mushroom can push up and lift the pier (courtesy Fu Hua Chen, Chen and Associates, Denver, Colorado).

of the current moisture condition with that reported at the time of initial site investigation or construction is also an important indicator of potential expansion.

A site investigation can often indicate the nature of the movement from the location and type of damage. Crack patterns can be evaluated with regard to the structural design, movement constraints, and the condition of the foundation soil. It is important to locate foundation cracks because they can severely affect structural integrity, and they will influence wall cracking and distress. Movements can also be transmitted to the roof framing, which should be checked. Crack patterns and magnitude of cracks indicate severity and cause of damage. However, care must be taken to evaluate relationships of crack patterns because not all cracking is due to soil volume changes and the cause of a particular crack is not always apparent.

Possible sources of water infiltration should be investigated. A rise in the groundwater table or presence of a perched water table should be observed. Utility lines and sprinkler systems should be pressure checked for leaks.

The type of drainage control, such as gutters and downspouts, gives an indication of possible problems due to surface water intrusion. The location of water spigots and shrubbery and the amount of irrigation done can also indicate possible causes of swelling. It should be noted that factors other than poor drainage control can contribute significantly to heaving, and repair of the drainage system alone usually will not solve the problem.

There is commonly an adjustment period for soil conditions after the remedial measures have been instituted. For example, changes in soil water content can occur for a period of time even if the treatment method is designed to prevent moisture changes due to surface water migration. Therefore, cosmetic repairs and even some structural repair of cracks in walls and foundations should be postponed until conditions in the foundation soils have stabilized. The adjustment period can range from several months to over a year depending on factors such as the method used, properties of the soil, depth of the active zone, and the soil moisture condition.

7.2.2   Remedial Procedure Alternatives

There are no standard procedures in the application of remedial measures. In general, innovation is necessary in the design of different systems. A complete listing of procedures that have been used would require the discussion of case studies too numerous to be included here. Some remedial measures will be discussed that can be used on particular foundation types or pavements. Some examples of the use of soil treatment techniques and currently used remedial procedures will also be included.

7.2.2.1   *Drilled Pier and Beam Foundation*

Distress to structures founded on drilled piers in expansive soil can result from a variety of causes. During the site investigation the potential for distress due to the following causes should be considered.

- Uplift from swelling soils through skin friction on piers.
- Uplift on base of grade beams or basement walls.
- Uplift or shrinkage under floor slab on fill.
- Improper pier design, e.g., pier diameter too large, pier length too short, or pier reinforcement inadequate.
- Improper construction of void space under grade beams.
- Improper pier construction, e.g., mushroom at top of pier (Figure 7.4), pier not anchored in stable bedrock, poor concrete quality, or void in pier shaft.
- Lateral pressure on foundation walls.

The remedial measures can attempt to repair or replace structural elements or correct improper design features. In some cases, merely excavating the void space beneath grade beams may be sufficient, whereas in others, replacement of supports along with a full program of underpinning may be needed. Depending on the nature of the cause of distress some factors that should be considered include the following:

- Excavate to the depth of the seasonal moisture variation around the piers affected and either loosen the soil, or backfill with material such as expanded vermiculite to reduce uplift skin friction.
- Underpin the structure with new piers to replace improperly placed, damaged, or short piers.
- Correct "mushroom" at the top of piers.

- Excavate void space beneath grade beam to allow soil heave from lifting beam off of piers. Determine required void depth from laboratory tests.
- If uplift on slab is causing movement of grade beam, cut slab loose from the grade beam. This will require approval of a structural engineer.
- If trees are a problem, remove them, prune the roots, and install a vertical lean concrete membrane below the lateral root zone.
- If climate is prone to conditions of low precipitation, a horizontal or vertical membrane can stabilize moisture conditions.
- Construct interceptor drains to prevent ponding. Make sure wood products are well ventilated to avoid rot.
- Slope ground surface and sidewalks away from the structure.
- Install sump pumps to remove perched water.

7.2.2.2 *Underpinning*

Underpinning of foundations consists of installing new foundation elements so as to provide additional support or to transfer the structural loads to different soil or rock strata. When used in conjunction with expansive soils, underpinning generally refers to the installation of piers that then extend to a stable soil zone. The new piers can be used to support the old foundation walls or grade beams if they have sufficient structural integrity or if they have been repaired by post tensioning or some other means. Alternatively new grade beams can be installed.

If access is possible piers can be drilled adjacent to the structure and then excavated laterally at the top to provide a vertical pier under the existing grade beam. Figure 7.5 shows piers being drilled using drilling machines designed to be used in areas where access is limited. Forms such as waxed cardboard tubes should be used to form the pier and aid in isolating it from uplift skin friction. Alternatively, piers can be drilled vertically adjacent to the old foundation and a bracket can be provided by framing the top of the pier so that it extends under the old grade beam.

Low head clearance drilling machines have also been developed that permit drilling of piers interior to basements. One such machine is shown in Figure 7.6.

Other innovative means for underpinning also have been developed. A number of patented systems are in use. In one system, the foundation is exposed at isolated points and a steel bracket is placed on the footing as shown in Figure 7.7. A steel pipe is inserted through a fitting on the steel bracket and jacked into the foundation soil using the footing as a reaction. Jacking proceeds until a penetration resistance of prescribed magnitude is observed. At that point the steel bracket is drilled and bolted to the pipe that has been jacked into the ground. This system is particularly useful in areas where soil shrinkage is also of concern and if sound bedrock exists at relatively shallow depth. In areas where uplift skin friction can develop on the pipes care must be taken to eliminate friction.

Whenever underpinning is used, its success depends on the ability to found the piers on stable soil or rock strata and to ensure that uplift skin friction will not

(a)

(b)

FIGURE 7.5. This residence was underpinned using specially designed augers (a) which can drill at an angle to get beneath eaves and (b) mounted on a small loader to allow drilling in narrow spaces (courtesy Jim Hemphill, Hemphill Corporation, Tulsa, Oklahoma).

FIGURE 7.6. A low head clearance drilling machine allows drilling in basements.

cause future movement. This is important and without it the underpinning can be considered only a temporary measure.

Furthermore, the grade beams or foundation walls spanning between points of underpinning must have sufficient strength to carry the loads that are transferred to them. The grade beams must be isolated from the expansive soils. Figure 7.8 shows an example where a number of plinths had been placed to provide support for the grade beams between piers. The uplift pressures were of such a magnitude that the concrete failed and the reinforcing bars buckled.

7.2.2.3   Slab-on-Grade Foundations

Heaving of slabs can affect stud walls, wall paneling, interior partitions, staircase walls, door frames, water lines, furnace ducts, shelves, and bookcases. Typical interior damages are shown in Figures 7.9 and 7.10.

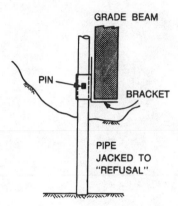

GRADE BEAM

PIN

BRACKET

PIPE
JACKED TO
"REFUSAL"

FIGURE 7.7. Underpinning foundation repair system.

Slab-on-grade foundations are intended to provide a rigid support that can withstand soil heave without resulting in large differential movements. Therefore, when damage occurs, the slab has not been sufficiently stiff. Remedial measures to correct this problem are difficult. Support to the slab at isolated points will probably be ineffective if the slab is not stiff enough to carry the loads between supports. Increasing the stiffness of the slab is difficult to accomplish at this point.

If access beneath the slab is possible, beams can be constructed under the slab and underpinned to provide additional support. If the load-bearing elements of the structure can be isolated, load can be transferred to outside piers or other forms of support. In some cases innovative techniques to increase the stiffness of the slab-on-grade by post tensioning the stiffened beams and/or superstructure have been successful. Frequently, however, it is necessary to remove the slab and construct a new foundation.

Remedial measures can also include structural adjustments, drainage control, and moisture control methods. The structural adjustments include isolating the structural elements from the soil by providing a void and/or nonswelling foundation soils beneath them. Slab replacement can be effective for floating slab construction when damage results only from heave of the slab between grade beams on stiffener beams. In this instance replacing the slab over granular fill, chemically stabilized soils, or using a structural wood floor over a crawl space can reduce future problems. Slab repairs should usually be accompanied with other remedial measures to prevent future volume changes in the foundation soils.

Drainage improvements can help to reduce further volume changes where damage is a result of an increase in the moisture content in foundation soils. Interceptor drains may be necessary if water intrusion is due to gravity flow of free water in a subsurface pervious layer. Surface drainage can be improved by regrading, altering the gutter and downspout system, and providing a moisture barrier or subdrain. A graded swale or ditch, preferably lined, can be constructed to divert surface water runoff away from the structure.

(a)

(b)

FIGURE 7.8. Uplift pressures on these plinths, placed under the grade beams between piers, were sufficient to crack the concrete and bend the reinforcing bars (courtesy U.S. Army Corps of Engineers, Fort Worth District).

7.2.2.4 Footing Foundations

Damage to footing and stem wall foundations usually is the result of the foundation walls not having sufficient strength and stiffness to withstand differential heave. If the foundations are shallow the methods discussed for slab on grade are also applicable. In this type of construction the structural loads are generally transferred to the foundation walls. Therefore jacking of the building and installation of a stiffer

(a)

(b)

FIGURE 7.9. Typical evidence of slab heave is shown by these damages to a high school building in Oklahoma: (a) distortion of a door frame and (b) uneven floors, as shown by the gap between these storage cabinets.

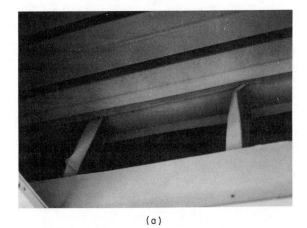

(a)

(b)

FIGURE 7.10. Interior damage caused by slab heave: (a) buckling of steel partition supports and (b) buckling of interior sheet rock partition walls.

foundation system are frequently possible. Underpinning of the new or stiffened old foundation can also aid in preventing additional heave.

A void is often recommended beneath part of the perimeter footing and slab after jacking so that additional heave will not cause further distress.

For basement walls, additional stiffness can be provided through post tensioning as shown in Figures 7.11 and 7.12. Alternatively a new grade beam can be constructed adjacent to the old wall as shown in Figures 7.13 and 7.14. This new grade beam is commonly referred to as a "sister wall."

Post tensioning, as shown in Figures 7.11 and 7.12 consists of placing cables on the inside and outside of the foundation walls. The cables are connected to steel brackets at the foundation corners and post tensioned to provide additional stiffness to the foundation walls. Locations of the cables and post-tensioning stresses should

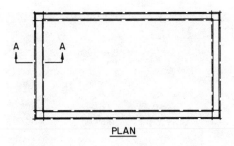

PLAN

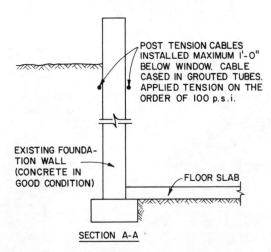

POST TENSION CABLES
INSTALLED MAXIMUM 1'-0"
BELOW WINDOW. CABLE
CASED IN GROUTED TUBES.
APPLIED TENSION ON THE
ORDER OF 100 p.s.i.

EXISTING FOUNDA-
TION WALL
(CONCRETE IN
GOOD CONDITION)

FLOOR SLAB

SECTION A-A

FIGURE 7.11. Post tensioning as a remedial measure (Chen, 1988).

FIGURE 7.12. Detail of the corner attachment for post tensioning cables.

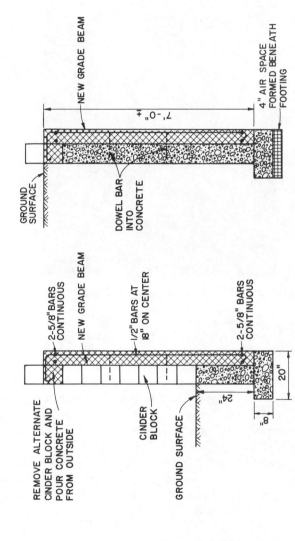

FIGURE 7.13. New grade beam or "sister wall" for repair of foundation walls (Chen, 1988).

FIGURE 7.14. Construction of a new grade beam.

be determined in consultation with a structural engineer and in consideration of the points of soil support and heave patterns.

It is important that cables be placed both on the inside and outside of the walls to prevent lateral buckling due to eccentric loading when the cables are post tensioned. Chen (1988) recommends casing the cable in grout-filled tubes to prevent corrosion.

Installation of a new grade beam or "sister wall" consists of casting a new reinforced concrete wall adjacent to the old one. The new grade beam should be well reinforced and tied into the old wall. The new grade beam should extend over the old wall as shown in Figure 7.13 to facilitate load transfer to the new one.

These techniques can be used alone or with underpinning to virtually replace the support provided by the old foundation. Cosmetic structural repair to the super-structure can be conducted after releveling.

7.2.2.5 Mud-Jacking and Injection

A remedial method that is often utilized in conjunction with underpinning of the foundation is mud-jacking of interior slabs. After a foundation has been leveled, there may be void spaces created beneath the slab. Mud-jacking consists of drilling holes through the floor slab and injecting a slurry below the slab to fill in voids, as illustrated in Figure 7.15. This procedure can be time consuming and messy and

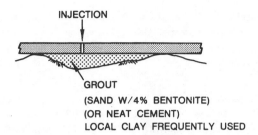

FIGURE 7.15. Mud-jacking is used to fill in void spaces and provide support beneath floor slabs.

generally requires that the interior of the structure be emptied of carpeting and interior furnishings.

If mud-jacking is used, care should be given to the design of the slurry that is utilized. In some cases a neat cement grout is used. Some clay fraction should be introduced into the grout such as bentonite in a concentration of approximately 4% by weight to provide for suspension of the cement until it can set. Frequently, local clay is used. However, if careful design is not utilized and highly plastic clays are used in too large a quantity, shrinking can occur as the grout dries, thereby reducing the effectiveness of the mud-jacking operation.

Another innovative technique that has been tried is foam injection. This is an experimental technique developed to overcome some of the problems associated with mud jacking. In this method, as shown in Figure 7.16, a foam mix is introduced through holes through the slab. Two chemicals are injected simultaneously that react to produce a foam that develops pressure beneath the slab and fills the void spaces. Lifting is controlled by the chemical mixture that defines the foam density. This procedure is still in developmental stages and requires further definition of injection intervals, and mix formulation to provide for appropriate swell pressures and other operational concerns.

7.2.2.6 Epoxy Treatment of Cracks

Slab foundations are designed to work as an integral unit, and major cracks may have a significant effect on the foundation–soil interaction. Injecting the cracks with epoxy can restore some structural integrity of the slab. This treatment is sometimes used together with mud-jacking to reduce differential movement of slabs and provide for cosmetic repair of finishes.

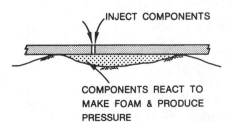

FIGURE 7.16. A proposed foam injection system alternative to mud-jacking.

The procedure involves the initial use of a quick set epoxy to seal the surface of the exposed crack. Ports are left at 18-in. intervals to inject epoxy into the crack after the surface has been sealed. This method is relatively easy to apply and generally proceeds rapidly.

7.2.2.7 Moisture Stabilization

A number of schemes have been reported in the literature that were designed to provide a means of stabilizing the water content in the foundation soils of houses and buildings. One system used in India relied on small trenches around the perimeter of a house, into which drainage water was directed. Schemes such as this, however, are subject to the same criticism as prewetting of foundation soils, primarily that it is difficult to assure uniform wetting to sufficient depth to be effective. Another method has been developed by Schmertmann (1983) in Gainesville, Florida, to be used in conjunction with remedial works.

In the Gainesville, Florida area, shrinking of soils is also of major concern as well as expansion. Schmertmann's technique is shown in Figure 7.17. Small diameter wells are drilled to a depth of 6 to 8 ft (1.8 to 2.4 m) along the interior of the foundation wall. Perforated PVC or plastic pipes are inserted in the wells and they are backfilled with sand and gravel. A header pipe is attached to all the wells and connected to a sump located near the foundation wall. The sump has a floating valve inserted in it to maintain a water level of approximately 1 or 2 ft (300 to 600 mm) above the header pipe. In this method, the water is introduced into the foundation soils over a continuous period of time and maintains them in a wet condition. The climate in that area is not arid or semiarid so drying of soils by evapotranspiration can be kept low by proper control of the vegetation around the structure. Large trees are removed to a distance of about 20 ft (6 m) from the foundation. Limitations regarding the use of this method in dry climates have been discussed in Section 5.3.1.

Personal communication with John Schmertmann (1983) indicates that this technique has been successful in reducing the differential settlement that had occurred previously in several foundations. In general it was indicated that the differential movement can be reduced by a factor of about two. The times required for this reduction in

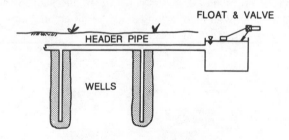

FIGURE 7.17. Moisture stabilization using the Schmertmann method.

settlement have ranged from as short as 6 months to as long as 2 to 3 years. At one research site, movements were monitored for a period of 2 years or more. Initial differential movement was indicated to be approximately 8 in. (200 mm) when the technique was implemented. Over the 2-year observation period this differential movement was reduced to approximately 4 in. (100 mm) by the above method. At that point the movement had been stabilized and structural and cosmetic repairs could be completed.

7.2.2.8 Moisture Barriers

Moisture barriers have been discussed in Chapter 6 as a means of controlling moisture variation in the foundation soil. In Chapter 6 moisture barriers were presented as a design feature but they can also be applied as part of a remedial action scheme. Several case histories have been reported in the literature where they have been implemented to minimize moisture content fluctuations sometimes in conjunction with moisture stabilization techniques as discussed in the previous section (Mohan and Rao, 1965; Woodward-Clyde-Sherard and Associates, 1968; Nadjer and Werno, 1973; Lee and Kocherhans, 1973; Poor, 1979).

Moisture barriers may be vertical or horizontal and should be designed in accordance with the discussion put forth in Chapter 6.

The time required for stabilization of foundation soil water contents using moisture barriers will be a function of the soil properties, the time for the water to redistribute by capillary action, and whether or not water is added. It may take several years for this to be accomplished. Cosmetic repairs should be postponed to the extent possible until an equilibrium condition is reached.

The successful application of vertical membranes coupled with the stabilization of moisture content by the addition of water in the foundation soil depends to a large extent on the experience and skill of the contractor. Sound judgment must be used to determine the amount of water to be applied and proper locations for adding it. Different types of foundation damage (e.g., cupped vs. domed slabs) would be considered differently.

It is important to recognize that the installation of barriers does not constitute a remedial measure in itself. Considerations of structural repairs must also be included. The barriers will aid in mitigation of future movement but the residual risk of movement after remedial action must be taken into account when advising clients.

7.3 REMEDIAL MEASURES FOR PAVEMENTS

Distress to pavements on expansive soils can usually be classified into one or more of the following four types (Kassiff et al., 1969)

- An unevenness along the length of the pavement with no visible cracking.
- Longitudinal cracking.
- Localized deformations near culverts or other structures, usually accompanied by lateral cracking.
- Localized failure of the pavement with disintegration of the road surface.

The elements of the preliminary investigation, as outlined for structural damages, are also applicable to highways, airfield pavements, railroads, canals, or other such transportation facilities. The site and soil investigation should consider all possible causes of distress so that remedial measures can be selected that will reduce future problems. The selection of remedial measures should be evaluated with respect to the cost of implementing the method, the amount of funds available, and the increase in serviceability that is expected. Most commonly the remedial measures comprise either removal and replacement, or construction of overlays. Whatever remedial measures are employed, care must be taken to ensure that previous design features that may have been responsible for the distress (e.g., poor drainage) are improved.

The selection of asphalt or concrete for the pavement material will depend on a number of factors, only one of which is the expansion potential of the subgrade. Each material has certain advantages.

If asphalt pavements are used on moderate to highly expansive subgrades, it is recommended that only full depth asphalt be used. Full depth asphalt has greater load-bearing potential than other designs, and it eliminates the need for granular base course which can provide access for water into the subgrade.

It is important when using full depth asphalt designs to also protect the shoulders from moisture variations. Longitudinal cracking near the shoulder can allow access for surface water to penetrate into the subgrade. It may be valuable to replace granular untreated bases with full depth asphalt bases in cases where damage occurred due to expansive soil subgrades. The use of a lime-treated gravel cushion may also prove effective and would be more economical than full depth asphalt.

Benefits of asphalt pavements include the following (Snethen, 1979a):

- Asphalt is flexible and some distortion is accommodated before failure occurs.
- The asphalt pavement provides a waterproof membrane to protect the subgrade.
- Asphalt pavement can be repaired relatively easily using specialized equipment such as a heater planner or removing and replacing damaged sections.

Reinforced Portland cement concrete pavements may also have particular advantages under some conditions. Recent advances in fiber reinforced concrete technology have led to significant improvements in the crack resistance and integrity of Portland cement concrete. Advantages of this material include the following:

- The rigid nature of the pavement can reduce differential heave if the expansion potential is low to moderate.
- Lifetime of the pavement may be longer.
- If curb and gutter are a part of the pavement, the use of asphalt can result in a joint between the curb and the pavement. If a Portland cement concrete pavement can be cast monolithically with the curb, that joint can be avoided. The elimination of the joint also eliminates a potential pathway for water to enter the subgrade.

The choice of full depth asphalt or reinforced concrete must be based on site-specific conditions and owner preference.

7.3.1 Remedial Maintenance

Remedial maintenance is a significant part of any state transportation agency's program, even though the methods often do not solve the recurring problems associated with expansive soil subgrades. Measures such as isolated overlays or patches to level distortions, mud-jacking, application of epoxy, and/or use of a heater planer can improve the ride quality for infrequent pavement distortions. They do not, however, reduce the swell potential of an expansive soil subgrade.

There are no particular guidelines established for the application of these repair procedures. They are usually only temporary in nature. For extensive distress caused by high-volume change soils, more extensive treatment alternatives will probably be necessary to prevent or reduce future problems.

For remedial measures to be effective the causes of the distress should be addressed. Ideally the appropriate design features should be implemented at the time of initial construction, but they also serve as important parts of the remedial program. In some cases, particularly for deeper soils, much of the swell that will eventually occur may have already taken place by the time of remedial action. The potential for future movement must be assessed carefully. The following techniques can be implemented to minimize volume change.

1. Remove and replace the problem soil.
2. Reduce the volume change characteristics of the clay by using additives.
3. Use a surcharge to confine the soil if swell pressures are not too great.
4. Minimize moisture changes in the subgrade.

Many of the above postconstruction treatment techniques can also be used as preconstruction alternatives. Since most of these methods have been discussed in detail in previous sections, the specifications and procedures involved will not be repeated. Some details specifically related to use in remedial applications will be reemphasized below.

7.3.2 Moisture Barriers

Horizontal moisture barriers, especially bituminous materials, are commonly applied using a variety of materials. Continuous sprayed asphalt, catalytically blown or emulsified, and asphalt-rubber can be effective when applied over the entire subgrade section (Snethen, 1979a). Pavement distortions should first be leveled and drainage adjustments made. Shoulders and ditches should be treated to a distance about 1.5 ft (500 mm) above the ditch invert up the back slope or to 0.5 to 1 ft (150 to 300 mm) above the elevation of the finished pavement grade. The Colorado Division of Highways (Brakey, 1973; Gerhardt, 1973) and the Arizona Department of Transportation (Fortsie et al., 1979) have achieved successful control of moisture variation in the subgrade using sprayed asphalt membranes. Longitudinal shrinkage cracking can be reduced if the membrane is extended several feet past the pavement edge and underlies the road shoulder (Williams, 1975).

For subgrades having a low swell potential, no soil treatment may be necessary when using a full depth asphalt pavement. Subgrades with marginal or high swell potential should first be treated with one of the previously discussed soil treatment methods (see Chapter 6) to reduce the likelihood for volume change before placing the full depth asphalt pavement. When using concrete pavement, an asphalt-treated base course should be placed continuously from shoulder to shoulder under the pavement.

7.3.3 Removal, Replacement, and Compaction Control

Usually subexcavation can be applied practically to a maximum depth of about 4 ft (1.2 m). In many instances the active zone of the expansive soil subgrade extends to a depth greater than 4 ft (1.2 m) and subexcavation and replacement alone will not prevent further volume changes. However, differential heave can be reduced.

Removal of the problem soil is usually a last resort technique. It should be used in conjunction with other appropriate design features.

The subexcavated soil should be replaced with nonexpansive fill compacted with careful attention to specified moisture and density criteria. Expansive subgrade that is not removed should be protected using moisture barriers, drainage improvement, and/or nonexpansive impermeable fill material. The depth of subexcavation required can be regulated on the basis of plasticity index or prediction of anticipated swell based on modified properties of the backfill and subgrade.

Subexcavation and remolding can also be an effective way to improve the characteristics of some problem materials. This has been discussed in detail in Chapter 6. When using restructured expansive material as fill, compaction criteria should be closely maintained. Lower dry densities and higher water contents produce lower swell potentials.

7.3.4 Drainage

Improper drainage is probably the most important factor contributing to soil volume change and subsequent damage to pavement. If water is allowed to stand in drainage ditches, it can penetrate the subgrade and amplify heave. Also, in many instances the pavement shoulders are untreated, providing an access for water infiltration. Moisture migration can occur in all directions due to capillary suction.

Signs indicating improper drainage include the existence of ponded water in ditches, soft spots in ditches or on slopes, or the presence of wetland plants such as cattails. The presence of sloughing at cut sections indicates a loss of soil strength, which may be related to drainage factors.

Periodic maintenance of drainage facilities is essential. Slough areas at cut sections should be removed, ponded water should be drained, and surface and subsurface drainage systems should be cleared of obstructions. Subsurface outlet markers should be maintained to assure that the outlet remains clear and drainage is provided away from the outlet.

Some state transportation agencies utilize standard design practices to ensure proper drainage. The verge slope should be greater than 12% and ditch inverts should be about 25 ft (7.6 m) from the shoulder's edge. Sometimes a minimum vertical distance between the pavement and the ditch invert is required. Minimum ditch or invert slopes can be established so that ditches are cleared quickly. As a remedial measure, design details for the highway in question may have to be adjusted to improve surface drainage if this is a cause of damage.

BIBLIOGRAPHY

Acker, W. L. (1974). Basic procedures for soil sampling and core drilling. Acker Drilling Co., Scranton, PA.

Acum, W. E. A., and Fox, L. (1951). Computation of load stresses in a three-layer elastic system. Geotechnique 2:293–300.

Aitchison, G. D. (1961). Relationship of moisture stress and effective stress functions in unsaturated soils. Pore Pressure and Suction in Soils. Butterworths, London, pp. 47–52.

Aitchison, G. D., and Martin, R. (1973). The quantitative description of the stress-deformation behavior of expansive soils. 2. A membrane oedometer for complex stress-path studies in expansive clays. Proc. 3rd Int. Conf. Expansive Soils, Haifa, Israel 2:83–88.

Aitchison, G. D. et al.—Review Panel (1965). Statement of the Review Panel. Moisture Equilibria and Moisture Changes in Soils Beneath Covered Areas. A Symposium in Print. Butterworths, Australia, pp. 7–21.

Altmeyer, W. T. (1955). Discussion of engineering properties of expansive clays. Proc. Am. Soc. Civil Eng. 81 (Separate No. 658):17–19.

American Association of State Highway and Transportation Officials (1978). Standard specifications for transportation materials and methods of sampling and testing, 12th Ed., Washington, D.C.

American Concrete Institute (1969). Recommended practice for concrete floor and slab construction. ACI Standard 302-69.

Anderson, J. U., Fadul, K. E., and O'Connor, G. A. (1972). Factors affecting the coefficient of linear extensibility in vertisols. Soil Sci. Soc. Am. Proc. 27:296–299.

ASTM (1970). Special Procedures for Testing Soil and Rock for Engineering Purposes. American Society for Testing and Materials, Special Technical Publication 479, 5th ed.

ASTM (1971). Sampling of Soil and Rock. American Society for Testing and Materials, Special Technical Publication 483.

ASTM Standards (1991). Natural Building Stones: Soil and Rock. Annual Book of ASTM Standards, Vol. 4, Philadelphia.

Aylmore, L. A. G., Quirk, J. P., and Sills, I. D. (1969). Effects of heating on the swelling of clay minerals. Highway Res. Board Special Rep. 103:31–38.

Baker, R., Kassiff, G., and Levy, A. (1973). Experience with a psychrometric technique. Proc. 3rd Int. Conf. Expansive Soils, Haifa, Israel 1:83–95.

Barden, L., Madedor, A. O., and Sides, G. R. (1969). Volume change characteristics of unsaturated clay. Soil Mech. Found. Div. ASCE 95 (No. SM1, Proc. Paper 6338, Jan.):33–52.

Barshad, I. (1965). Thermal analysis techniques for mineral identification and mineralogical composition. Methods of Soil Analysis, American Society of Agronomy Monograph, No. 9, Ch. 50.

Basu, R., and Arulanandan, K. (1973). A new approach to the identification of swell potential in soils. Proc. 3rd Int. Conf. Expansive Soils, Haifa, Israel 1:1–11.

Benson, J. R. (1959). Discussion of expansive clays—properties and problems by W. G. Holtz. Quart. Colorado School Mines 54(4):117–124.

Bishop, A. W. (1959). The principle of effective stress. Technisk Ukeflad, No. 39.

Bishop, A. W., and Blight, G. E. (1963). Some aspects of effective stress in saturated and unsaturated Soils. Geotechnique 13:177–197.

Bishop, A. W., and Henkel, D. J. (1962). The Measurement of Soil Properties in the Triaxial Test, 2nd ed. William and Clowes, London.

Bjerrum, L., Nash, J. K. T. L., Kennard, R. M., and Gibson, R. R. (1972). Hydraulic fracturing in field permeability testing. Geotechnique 22(2):319–332.

Blight, G. E., and deWet, J. A. (1965). The acceleration of heave by flooding. Moisture Equilibria and Moisture Changes Beneath Covered Areas. A Symposium in Print. Butterworths, Australia, pp. 89–92.

Bocking, K. A., and Fredlund, D. G. (1980). Limitations of the axis translation technique. Proc. 4th Int. Conf. Expansive Soils, ASCE, Denver, CO 117–135.

Bohn, H. L., McNeal, B. L., and O'Conner, G. A. (1985). Soil Chemistry, 2nd ed. John Wiley, New York.

Brakey, B. A. (1973). Moisture stabilization by membranes, encapsulation and full depth paving. Proc. Workshop on Expansive Clays and Shales in Highway Design and Construction, University of Wyoming, Laramie 2:155–189.

Brakey, B. A. (1977). Presentation at Federally Coordinated Program Review Session. Federal Highway Administration. Atlanta, GA.

Brasher, B. R., Franzmeier, D. P., Valassis, V., and Davidson, S. E. (1966). Use of saran resin to coat natural soil clods for bulk density and water retention measurements. Soil Sci. 101:108.

Brown, R. W., and van Haveren, B. P., eds. (1972). Psychrometry in water relations research. Utah Agric. Exp. Sta., Utah State Univ., Logan.

Buckley, E. L. (1974). Loss and damage on residential slab-on-ground foundations. Construction Research Center, Dept. Civil Eng., Univ. Texas Arlington, Rep. TR-2-74, 23 pp.

Building Research Advisory Board (BRAB). (1968). Criteria for selection and design of residential slabs-on-ground. Publ. 1571, National Academy of Sciences Rep. No. 33 to Federal Housing Administration, NTIS No. PB-261 551.

Burland, J. B. (1962). The estimation of field effective stresses and the prediction of total heave using a revised method of analyzing the double oedometer test. The Civil Engineer in South Africa, Tran. South African Inst. Civil Eng., July.

Burland, J. B. (1965). Some aspects of the mechanical behavior of partly saturated soils. Moisture Equilibria and Moisture Changes in Soils Beneath Covered Areas. A Symposium in Print. Butterworths, Australia, pp. 270–278.

Burmister, D. M. (1943). The theory of stresses and displacements in layered systems and application to the design of airport runways. Proc. Highway Res. Board.

Burmister, D. M. (1945). The general theory of stresses and displacements in layered soil systems. J. Appl. Phys. 16.

Burmister, D. M. (1958). Evaluation of pavement systems of the AASHO road test by layered systems methods. Highway Res. Board Bull. 177.

Campbell, G. S., and Gardner, W. H. (1971). Psychrometric measurement of soil water potential: Temperature and bulk density effects. Soil Sci. Soc. Proc. 35:8–12.

Casagrande, A. (1936). The determination of the pre-consolidation load and its practical significance. Discussion D-34, Proc. 1st Int. Conf. Soil Mech. Found. Eng., Cambridge 3:60–64.

Casagrande, L. (1960). Practical aspects of electroosmosis in foundation engineering. First Pan-American Conference on Soil Mechanics Foundation Engineering, Vol. 1, Mexico.

Chapman, H. D. (1965). Cation-exchange capacity. Methods of Soil Analysis. American Society of Agronomy Monograph, No. 9, Ch. 57.

Chen, F. H. (1965). The use of piers to prevent the uplifting of lightly loaded structures founded on expansive soil. Concluding Proc. Eng. Effects of Moisture Changes in Soils, Int. Res. Eng. Conf. Expansive Clay Soils, Supplementing the Symposium in Print, Texas A&M Press, pp. 152–171.

Chen, F. H. (1973). The basic physical property of expansive soils. Proc. 3rd Int. Conf. Expansive Soils, Haifa, Israel 1:17–25.

Chen, F. H. (1988). Foundations on Expansive Soils. American Elsevier Science Publ., New York.

Coleman, J. D. (1962). Stress/strain relations for partly saturated soils. Correspondence. Geotechnique 12(4):348–350.

Colorado Department of Highways and University of Colorado. (1964). A review of literature on swelling soils. Colorado State Department of Highways, 65 pp.

Compton, C. P. V. (1970). A study of the swelling behavior of an expansive clay as influenced by the clay microstructure, soil suction, and external loading. Ph.D. thesis, Texas A&M University, College Station, TX.

Croney, D., Coleman, J. D., and Black, W. P. M. (1958). Movement and distribution of water in soil in relation to highway design and performance. National Research Council, Highway Research Board, Special Rep. 40, Washington, D.C., pp. 226–252.

Currin, D. D., Allen, J. J., and Little, D. N. (1976). Validation of Soil Stabilization Index System with Manual Development. U.S. Air Force Tech. Rep. SRL-TR-76-0006.

Davidson, J. L. (1983). Cambridge self-boring pressure meter. Short Course on New Methods of in Situ Testing, Orlando, FL, March.

Davidson, L. K., Demirel, T., and Handy, R. L. (1965). Soil pulverization and lime migration in soil-lime stabilization. Highway Res. Rec. 92:103–126.

Dawson, R. F. (1953). Movement of small houses erected on an expansive clay soil. Proc. 3rd Int. Conf. Soil Mech. Found. Eng. 1:346–350.

Dawson, R. F. (1959). Modern practices used in the design of foundations for structures on expansive soils. Quart. School Mines 54(4):66–87.

De Bruijn, C. M. A. (1961). Swelling characteristics of a transported soil profile at Leeuhof Vereeniging (Transvaal). Proc. 5th Int. Conf. Soil Mech. Found. Eng. 1:43–49.

De Bruijn, C. M. A. (1965). Some observations on soil moisture conditions beneath and adjacent to tarred roads and other surface treatments in South Africa. Moisture Equilibria and Moisture Changes Beneath Covered Areas. A Symposium in Print, Butterworths, Australia: 135–142.

De Bruijn, C. M. A. (1973). Moisture redistribution in southern African soils. 8th Int. Conf. Soil Mech. Found. Eng., Moscow 8:37–44.

Diller, D. G. (1973). Expansive soils in Wyoming highways. Pp. 250–255 in D. R. Lamb and S. J. Hanna, eds. Proceedings of Workshop on Expansive Clays and Shales in Highway Design and Construction, Vol. 2. Federal Highway Administration. Denver, CO. 304 pp.

Donaldson, G. W. (1965). A study of level observations on buildings as indications of moisture movements in underlying soils. Moisture Equilibria and Moisture Changes Beneath Covered Areas. A Symposium in Print, Butterworths, Australia: 156–164.

Donaldson, G. W. (1973). The prediction of differential movement on expansive soils. Proc. 3rd Int. Conf. Expansive Soils, Haifa, Israel, 1:289–293.

Dowding, C. H. (1979). Perspectives and challenges of site characterization. Proc. ASCE Speciality Workshop Site Characterization, Evanston, Illinois: 10–35.

Drumright, E. E. (1989). The contribution of matric suction to the shear strength of unsaturated soils. Ph.D. dissertation, Colorado State University, Fort Collins, CO.

Eads, J. L. and Grim, P. E. (1966). A quick test to determine lime requirements for lime stabilization. Highway Res. Rec. 139:61–72.

Edgar, T. V., Nelson, J. D., and McWhorter, D. B. (1989). Nonisothermal consolidation of unsaturated soil. J. Geotechn. Eng. Div., ASCE 115(10:1351–1372, October).

Escario, V., and Saez, L. (1973). Measurement of the properties of swelling and collapsing soils under controlled suction. Proc. 3rd Int. Conf. Expansive Soils, Haifa, Israel v.1:195–200.

Fargher, P. J., Woodburn, J. A., and Selby, J., eds. (1979). Footings and foundations for small buildings in arid climates, with special reference to South Australia. The Institution of Engineers, Australia, S. Australian Division.

Fawcett, R. C., and Collis-George, N. (1967). A filter paper method of determining the moisture characteristics of soil. Aust. J. Exp. Agric. Animal Husbandry 7:162–167.

Federal Aviation Administration. (1967). Airport paving. AC150/5320-6H, U.S. Dept. of Transportation.

Felt, E. J. (1953). Influence of vegetation on soil moisture contents and resulting soil volume changes. Proc. 3rd Int. Conf. Soil Mech. Found. Eng. 1:24–27.

FEMA (1982). Special statistical summary—data, injuries and property loss by type of disaster 1970–1980. Federal Emergency Management Agency, April.

Fohs, D. G., and Kinter, E. P. (1972). Migration of lime in compacted soil. Public Roads 37(1):1–8.

Fortsie, D., Walsh, H., and Way, G. (1979). Control of expansive clays under existing highways. Proc. 16th Paving Conf., University of New Mexico, Albuquerque: 13–33.

Fox, I.. (1948). Computation of traffic stresses in a simple road structure. Dept. of Scientific and Industrial Research, Road Research Tech. Paper 9.

Franzmeier, D. P., and Ross, S. J. (1968). Soil swelling: Laboratory measurement and relation to other soil properties. Soil Sci. Soc. Am. Proc. 32:573–577.

Fraser, R. A., and Wardle, L. J. (1975). The analysis of stiffened raft foundations on expansive soils. Proc. Symp. Recent Developments in the Analysis of Soil Behavior and Their Application to Geotechnical Structures, Univ. of N.S.W., Sydney, Australia: 89–98.

Fredlund, D. G. (1969). Consolidometer test procedural factors affecting swell properties. Proc. 2nd Conf. Expansive Clay Soils. Texas A&M Univ. Press, College Station, TX, pp. 435–456.

Fredlund, D. G. (1973). Volume change behavior of unsaturated soils. Ph.D. dissertation, University of Alberta, Edmonton.

Fredlund, D. G. (1979). Appropriate concepts and technology for unsaturated soils. 2nd Canadian Geotech. Colloquium, Canadian Geotech. J., V.16, No. 1:121–139.

Fredlund, D. G. (1983). Prediction of ground movements in swelling clays. 31st Annu. Soil Mech. Found. Eng. Conf., Univ. of Minnesota, Minneapolis, February.

Fredlund, D. G., and Morgenstern, N. R. (1973). Pore pressure response below high air entry discs. Proc. 3rd Int. Conf. Expansive Soils, Haifa, Israel I:97–108.

Fredlund, D. G., and Morgenstern, N. R. (1977). Stress state variables for unsaturated soils. J. Geotech. Eng. Div., ASCE 103 (GT5):447–466.

Fredlund, D. G., Hasan, J. U., and Filson, H. L. (1980). The prediction of total heave. Proc. 4th Int. Conf. Expansive Soils, Denver, CO: 1–17.

Gardner, R. (1937). A method for measuring the capillary tension of soil moisture over a wide moisture range. Soil Sci. 43:277–283.

Gardner, W. H. (1965). Water content. Methods of Soil Analysis, C. A. Black, D. D. Evans, J. L. White, L. E. Ensminger and F. E. Clark, eds. Agronomy 9:82–127.

Gerhardt, B. B. (1973). Soil modification highway projects in Colorado. Proc. Workshop Expansive Clays and Shales in Highway Design and Construction, Univ. of Wyoming, Laramie 2:33–48.

Gibbs, H. J. (1973). Use of a consolidometer for measuring expansion potential of soils. Proc. Workshop Expansive Clays and Shales in Highway Design and Construction. Univ. Wyoming, Laramie, May 1:206–213.

Gizienski, S. F., and Lee, L. J. (1965). Comparison of laboratory swell tests to small scale field tests. Engineering Effects of Moisture Changes in Soils, Concluding Proceedings. International Research and Engineering Conference on Expansive Clay Soils. Texas A&M Press, College Station, TX, pp. 108–119.

Gokhale, K. V. G. K., and Swaminathan, E. (1973). Accelerated stabilization of expansive soils. Proc. 3rd Int. Conf. Expansive Soils, Divisions 1–5, Haifa, Israel 1:35–41.

Goode, J. C. (1982). Heave prediction and moisture migration beneath slabs on expansive soils. M.S. thesis, Colorado State University, Fort Collins, CO.

Grim, R. E. (1968). Clay Mineralogy, 2nd ed. McGraw-Hill, New York.

Grisak, J. E., and Cherry, J. A. (1975). Hydraulic characteristics and response of fractured till and clay confining a shallow aquifer. Can. Geotech. J. 12(1):23–43.

Gromko, G. J. (1974). Review of expansive soils. J. Geotech. Eng. Div., ASCE 100 (GT6):667–687.

Grossman, R. B., Brashner, B. R., Franzmeier, D. P., and Walker, J. L. (1968). Linear extensibility as calculated from natural-clod bulk density measurements. Soil Sci. Soc. Am. Proc. 32(4):570–573.

Hamberg, D. J. (1985). A simplified method for predicting heave in expansive soils. M.S. thesis, Colorado State University, Fort Collins, CO.

Hamberg, D. J., and Nelson, J. D. (1984). Prediction of floor slab heave. 5th Int. Conf. Expansive Soils, Adelaide, S. Australia. 137–217.

Hamilton, J. J. (1969). Effects of environment on the performance of shallow foundations. Canadian Geotech. J. 6:65–80.

Hammitt, G. M., II, and Ahlvin, R. G. (1973). Membranes and encapsulation of soils for control of swelling. Proceedings of Workshop on Expansive Clays and Shales in Highway Design and Construction, D. R. Lamb and S. J. Hanna, eds. Federal Highway Administration, Denver, CO, pp. 80–95.

Hardy, J. R. (1971). Factors influencing the lime reactivity of tropically and subtropically weathered soils. Ph.D. thesis, University of Illinois at Urbana.

Hart, S. S. (1974). Potentially swelling soil and rock in the front range urban corridor, Colorado. Environmental Geology 7, Colorado Geological Survey, Denver, CO.

Hartronft, B. C., Buie, L. D., and Hicks, F. P. (1969). A study of lime treatment of subgrade to depths of two feet. Research and Development Division, Oklahoma Highway Department, Oklahoma City.

Haynes, J. H., and Mason, R. C. (1965). Subgrade soil treatment at the Apparel Mart—Dallas, Texas. Engineering Effects of Moisture Changes in Soils. Concluding Proceedings, International Research and Engineering Conference on Expansive Soils. Texas A&M Press, College Station, TX, pp. 172–182.

Higgins, C. M. (1965). High pressure lime injection. Louisiana Dept. of Highways, Research Report 17, Interim Report 2, August.

Hilf, J. W. (1956). An investigation of pore-water pressure in compacted cohesive soils. Ph.D. thesis. Tech. Memo. No. 654, U.S. Dept. Interior Bureau of Reclamation, Design and Construction Div., Denver, CO.

Holland, J. E., and Lawrence, C. E. (1980). Seasonal heave of Australian clay soil. Proc. 4th Int. Conf. Expansive Soils, Denver, CO 1:302–321.

Holland, J. E., and Richards, J. (1984). The practical design of foundations for light structures on expansive clays. Proc. 5th Int. Conf. Expansive Soils, Adelaide, S. Australia 154–158.

Holland, J. E., Pitt, W. G., Lawrence, C. E., and Climino, D. J. (1980). The behavior and design of housing slabs on expansive clays. Proc. 4th Int. Conf. Expansive Soils, Denver, CO V.1:448–468.

Holt, J. H. (1969). A study of physico-chemical, mineralogical, and engineering index properties of fine-grained soils in relation to their expansive characteristics. Ph.D. Dissertation, Texas A&M Univ., College Station, TX.

Holtz, W. G. (1959). Expansive clays—properties and problems. Quart. Colorado School Mines 54(4):89–117.

Holtz, W. G., and Gibbs, H. J. (1956). Engineering properties of expansive clays. Transact. ASCE 121:641–677.

Howard, A. K. (1977). Laboratory classification of soils—Unified Soil Classification System. Earth Sciences Training Manual No. 4, U.S. Bureau of Reclamation, Denver. 56 pp.

Hvorslev, M. J. (1948). Subsurface exploration and sampling of soils for engineering purposes. U.S. Army Corps of Eng., Waterways Exp. Sta., Vicksburg, MS, 521 pp; reprinted by the Engineering Foundation, 1962.

Ingles, O. G., and Neil, R. C. (1970). Lime grout penetration and associated moisture movements in soil. CSIRO, Australia, Division of Applied Geomechanics, Research Paper 138.

Jennings, J. E. B. (1961). A revised effective stress law for use in the prediction of the behavior of unsaturated soils. Pore Pressure and Suction in Soils. Butterworths, London, pp. 26–30.

Jennings, J. E. B., and Burland, J. B. (1962). Limitations to the use of effective stress in partly saturated soils. Geotechnique 12(2):125–144.

Jennings, J. E. B., and Kerrich (1962). The heaving of buildings and the associated economic consequences with particular reference to the orange free state goldfields. The Civil Engineer in South Africa. Transact. S. African Inst. Civil Eng., November. V.4, No.11:221–248.

Jennings, J. E. B., and Knight, K. (1957). The prediction of total heave from the double oedometer test. Transact. S. African Inst. Civil Eng. 7:285–291.

Jennings, J. E. B., Firtu, R. A., Ralph, T. K., and Nagar, N. (1973). An improved method for predicting heave using the oedometer test. Proc. 3rd Int. Conf. Expansive Soils, Haifa, Israel 2:149–154.

Jezequel, J. F., and Mieussens, C. (1975). In situ measurement of coefficients of permeability and consolidation in fine soils. In Situ Measurement of Soil Properties, Specialty Conference of the Geotechnical Eng. Div., ASCE, North Carolina State Univ., June 1975 I:208–224.

Jobes, W. P., and Stroman, W. R. (1974). Structures on expansive soils. Tech. Rep. M-81, Construction Engineering Research Lab, Champaign, IL, April.

Johnson, L. D. (1969). Review of literature on expansive clay soils. U.S. Army Eng. Waterways Exp. Sta., Vicksburg, MS, Misc. Paper S-73-17.

Johnson, L. D. (1973). Influence of suction on heave of expansive soils. U.S. Army Eng. Waterways Exp. Sta., Vicksburg, MS, Misc. Paper S-73-17.

Johnson, L. D. (1977). Evaluation of laboratory suction tests for prediction of heave in foundation soils. Army Eng. Waterways Exp. Station, Vicksburg, MS, Rep. WES-TR-S-77-7, August.

Johnson, L. D. (1978). Predicting potential heave and heave with time in swelling soils. U.S. Army Eng. Waterways Exp. Sta., Vicksburg, MS, Tech. Rep. S-78-7.

Johnson, L. D. (1979). Overview for design of foundations on expansive soils. U.S. Army Eng. Waterways Exp. Sta., Vicksburg, MS, Misc. Paper GL-79-21.

Johnson, L. D. (1980). Field test sections on expansive soil. Proc. 4th Int. Conf. Expansive Soils, Denver, CO 1:262–283.

Johnson, L. D. (1981). Field test sections on expansive soils. U.S. Army Eng. Waterways Exp. Sta., Vicksburg, MS, Rep. 6L-81-4, May.

Johnson, L. D., and McAnear, C. L. (1973). Controlled field tests of expansive soils. Proc. Workshop on Expansive Clays and Shales in Highway Design and Construction, Univ. of Wyoming, Laramie 1:137–159.

Johnson, L. D., and McAnear, C. L. (1974). Controlled field tests of expansive soils. Bull. Assoc. Eng. Geologists XI(4):353–369.

Johnson, L. D., and Snethen, D. R. (1978). Prediction of potential heave of swelling soil. Geotech. Testing J. 1(3):117–124.

Johnson, L. D., and Stroman, W. R. (1976). Analysis of behavior of expansive soil foundations. Army Eng. Waterways Exp. Sta., Vicksburg, MS, Rept. No. WES-TR-S-76-8, June.

Johnson, L. D., Sherman, W. C., and McAnear, C. L. (1973). Field test sections on expansive clays. Proc. 3rd Int. Conf. Expansive Soils, Haifa, Israel V.1:239–248.

Johnson, S. J. (1958). Cement and clay grouting of foundations: Grouting with clay-cement grouts. ASCE J. Soil Mech. Found. Div. 884 (SM1), Paper 1545. 12 pp.

Jones, A. (1962). Tables of stresses in three-layer elastic systems. Highway Res. Board Bull. 342.

Jones, C. W. (1958). Stabilization of expansive clay with hydrated lime and with Portland cement. Highway Res. Board Bull. 193:40–47.

Jones, D. E. (1981). Perspectives on needs for an availability of scientific and technical information. Dept. Housing and Urban Development, presented at 1st meeting of Committee on Emergency Management, Commission on Sociotechnical Systems, National Research Council, April 30–May 1, 1981, Washington, D.C.

Jones, D. E., and Holtz, W. G. (1973). Expansive soils—the hidden disaster. Civil Eng., ASCE 43(8):49–51.

Kassiff, G. and Baker, R. (1971). Aging effects on swell potential of compacted clay. J. Soil Mechanics and Foundation Div., ASCE, SM 3:529–540.

Kassiff, G., and Wiseman, G. (1966). Control of moisture and volume change in clay subgrades by subdrainage. Highway Res. Board Rec. 111:1–11.

Kassiff, G., Livneh, M., and Wiseman, G. (1969). Pavements on Expansive Clays. Jerusalem Academic Press, Jerusalam, Israel.

Kelly, J. E. (1973). Lime stabilization of expansive clays at the Dallas–Fort Worth Airport and movie commentary. Proc. of Workshop on Expansive Clays and Shales in Highway Design and Construction, Vol. 2, D. R. Lamb and S. J. Hanna, eds. Federal Highway Administration, Denver, CO, pp. 20–32.

Komornik, A., Wiseman, G., and Ben-Yaacob, Y. (1969). Studies of in situ moisture and swelling potential profiles. Proc. 2nd Int. Res. Eng. Conf. Expansive Clay Soils, Texas A&M Univ., College Station, TX, pp. 348–361.

Komurka, V. E. (1985). Settlement and drainage of uranium mill tailings. M.S. thesis, Colorado State University, Fort Collins, CO.

Krahn, J., and Fredlund, D. G. (1971). On total, matric and osmotic suction. Soil Sci. 114(5):339–345.

Krazynski, L. M. (1976). Engineering Properties of Expansive Soils. Geological Soc. of America, Eng. Geology Div., Symp. Expansive Soils, Denver, CO.

Krazynski, L. M. (1979). Site investigation. The design and construction of residential slabs-on-ground: State-of-the-art. Proc. Workshop Building Research Advisory Board (BRAB), Commission of Sociotechnical Systems, Washington, D.C. I:33–48.

Krazynski, L. M. (1980). Expansive soils in highway construction—some problems and solutions. 4th Int. Road Fed. African Highways Conf., Nairobi, Kenya, January.

Krohn, J. P., and Slossan, J. E. (1980). Assessment of expansive soils in the United States. Proc. 4th Int Conf. Expansive Soils, Denver, CO 1:596–608.

Lambe, T. W. (1960a). A mechanistic picture of shear strength in clay. Proc. ASCE Res. Conf. Shear Strength of Cohesive Soils, Univ. of Colorado, Boulder, CO 555–580.

Lambe, T. W. (1960b). The character and identification of expansive soils, soil PVC meter. Federal Housing Administration, Technical Studies Program, FHA 701.

Lambe, T. W., and Whitman, R. V. (1969). Soil Mechanics. John Wiley, New York.

Lee, L. J., and J. G. Kocherhans. (1973). Soil stabilization by use of moisture barriers. Proc. 3rd Int. Conf. Expansive Soils, Haifa, Israel 1:295–301.

Lee, R. K. C., and Fredlund, D. G. (1984). Measurement of soil suction using the MCS 6000 sensor. Proc. 6th Int. Conf. Expansive Soils, Adelaide, S. Australia 50–54.

Lundy, H. L., Jr. and Greenfield, B. J. (1968). Evaluation of deep in situ soil stabilization by high pressure lime slurry injection. Highway Res. Rec. 235:27–35.

Lytton, R. L. (1969). Theory of moisture movement in expansive clays. Research Rep. 118-1, Center for Highway Research, Univ. of Texas at Austin. 161 pp.

Lytton, R. L. (1970). Design criteria for residential slabs and grillage rafts on reactive clay. Rept. for Australian Commonwealth Scientific and Industrial Research Organization C.S.I.R.O., Melbourne, Australia, November.

Lytton, R. L. (1972). Design method for concrete mats on unstable soils. Third Inter-American Conf. Materials Tech., Rio de Janiero, Brazil 171–177.

Lytton, R. L. (1973). Expansive clay roughness in the highway design system. Proc. Workshop Expansive Clays and Shales in Highway Design and Construction, Univ. of Wyoming, Laramie 2:123–149.

Lytton, R. L., and Woodburn, J. A. (1973). Design and performance of mat foundation on expansive clay. Proc. 3rd Int. Conf. Expansive Soils, Haifa, Israel 1:301–308.

Madrid, L. D. (1984). Compressibility and shear strength of unsaturated spent soil shale. Masters thesis, Colorado State University, Fort Collins, CO.

Marchetti, S. (1975). A new in situ test for the measurement of horizontal soil deformability. Proc. Conf. In Situ Measurement of Soil Properties, ASCE Specialty Conf., Raleigh, NC, 2:255–259.

Marchetti, S. (1980). In situ tests by flat dilatometer. J. Geotech. Eng. Div. ASCE, No. GT3, March 299–830.

Marshall, T. J., and Holmes, J. W. (1979). Soil Physics. Cambridge University Press, Cambridge.

Mateos, M. (1964). Soil lime research at Iowa State University. ASCE J. Soil Mech. Found. Div. 90 (SM2):127–153.

Matyas, E. L., and Radhakrishna, H. S. (1968). Volume change characteristics of partially saturated soils. Geotechnique 18(4):432–448.

McCormick, D. E., and Wilding, L. P. (1975). Soil properties influencing swelling in Canfield and Geeburg soils. Soil Sci. Soc. Am. Proc. 39:496–502.

McCrone, W. C., and Delly, J. G. (1973). The Particle Atlas, 2nd ed., Vol. 1, Principles and Techniques. Ann Arbor Science, Ann Arbor, MI.

McDonald, E. B. (1973). Experimental moisture barrier and waterproof surface. South Dakota Department of Transportation, Final Report HR0200 (3645).

McDonald, E. B., and Potter, A. W. (1973). Review of highway design and construction through expansive soils in South Dakota on I-95—Missouri River West—135 miles. Proc. Workshop Expansive Clays and Shales in Highway Design and Construction, Univ. of Wyoming, Laramie 2:230–244.

McDowell, C. (1959). The relation of laboratory testing to design for pavements and structures on expansive soils. Quart. Colorado School of Mines 54(4):127–153.

McDowell, C. (1965). Remedial procedures used in the reduction of detrimental effects of swelling soils. Engineering Effects of Moisture Changes in Soils. Concluding Proceedings, International Research and Engineering Conference on Expansive Clay Soils. Texas A&M Univ. Press, College Station, TX, pp. 239–275.

McKeen, R. G. (1976). Design and construction of airport pavements on expansive soils. Civil Engineering Research Institute, Univ. of New Mexico, Albuquerque, Rep. No. CERF-AP-18, June. 178 pp.

McKeen, R. G. (1977). Pavement design for swelling soils: A review. Proc. 14th Pavement Conf., University of New Mexico 14:88–131.

McKeen, R. G. (1980). Field studies of airport pavements on expansive clay. Fourth Int. Conf. Expansive Soils, Denver, CO 1:242–261.

McKeen, R. G. (1981). Design of airport pavements for expansive soils. U.S. Dept. of Transportation, Federal Aviation Administration, Rep. No. DOT/FAA/RD-81/25.

McKeen, R. G., and Hamberg, D. J. (1981). Characterization of expansive soils. Trans. Res. Rec. 790, Trans. Res. Board 73–78.

McKeen, R. G., and Johnson, L. D. (1990). Climate-controlled soil design parameters for mat foundations. J. Geotech. Eng. ASCE 116(7):1073–1094. July.

McKeen, R. G., and Lenke, L. R. (1982). Thickness design for airport pavements on expansive soils. New Mexico Engineering Research Institute (NMERI), Univ. of New Mexico, Albuquerque, NM.

McKeen, R. G., and Nielsen, J. P. (1978). Characterization of expansive soils for airport pavement design. U.S. Dept. of Transportation, Federal Aviation Administration, Rept. No. FAA-120-78-59.

McQueen, I. S., and Miller, R. F. (1968). Calibration and evaluation of wide-range gravimetric method for measuring soil moisture stress. Soil Sci. 10(3, June):521–527.

McQueen, I. S., and Miller, R. F. (1974). Approximating soil moisture characteristics from limited data: Empirical evidence and tentative model. Water Resources Res. 10(3, June):521–527.

Meyer, K. T., and Lytton, R. L. (1966). Foundations design in swelling clays. Presented at October Texas Section ASCE Meeting, Austin, TX, p. 52.

Meyn, R. L., and White, R. S. (1972). Calibration of thermocouple psychrometers: A suggested procedure for development of a reliable predictive model. Psychrometry in Water Relations Research, Utah Agricultural Experimental Station, Utah State Univ., Logan, pp. 56–64.

M.I.T. Report. (1963). Engineering behavior of partially saturated soils. M.I.T. Phase Report No. 1, soil stabilization.

Mitchell, J. K. (1976). Fundamentals of Soil Behavior. John Wiley, New York.

Mitchell, J. K. (1979). In situ techniques for site characterization. Proc. ASCE Specialty Workshop Site Charact. Exp., Evanston, IL 107–129.

Mitchell, J. K., and Raad, L. (1973). Control of volume changes in expansive earth materials. Proc. Workshop Expansive Clays and Shales in Highway Design and Construction, Vol. 2, D. R. Lamb and S. J. Hanna, ed. Federal Highway Administration, Denver, CO, pp. 200–219.

Mitchell, P. W. (1980). The concepts defining the rate of swell in expansive soils. Proc. 4th Int. Conf. Expansive Soils, Denver, CO 1:106–116.

Mohan, D., and Rao, B. G. (1965). Moisture variation and performance of foundations in black cotton soils in India. Moisture Equilibria and Moisture Changes in Soils Beneath Covered Areas. Symposium, Soil Mechanics Section, Commonwealth Scientific and Industrial Research Organization. Butterworths, Australia, pp. 175–183.

Moore, J. C., and Jones, R. L. (1971). Effect of soil surface area and extractable silica, alumina and iron on lime stabilization characteristics of Illinois soils. Highway Res. Rec. 351:87–92.

Morris, G. R. (1973). Arizona's experience with swelling clays and shales. Proc. Workshop Expansive Clays and Shales in Highway Design and Construction, Vol. 2, D. R. Lamb and S. J. Hanna, eds. Federal Highway Administration, Denver, CO, pp. 283–285.

Nadjer, J., and Werno, M. (1973). Protection of buildings on expansive clays. Proc. 3rd Int. Conf. Expansive Soils, Haifa, Israel 1:325–334.

National Soil Survey Laboratory. (1981). Principles and procedures for using soil survey laboratory data. Unpublished training materials, NSSC, SCS.

Navy, Dept. of, Naval Facilities Engineering Command. (1971). Design manual—Soil mechanics, foundations and earth structures. U.S. Naval Publications and Forms Center, NAVFAC DM-7.

Nelson, J. D., and Edgar, T. V. (1978). Moisture migration beneath impermeable membranes. Proc. 15th Annu. Symp. Eng. Geol. Soil Eng., Idaho Dept. of Highways, Boise, Idaho, April.

Newland, P. L. (1965). The behaviour of a house on a reactive soil protected by a plastic film moisture barrier. Engineering Effects of Moisture Changes in Soils. Proceedings,

International Research and Engineering Conference on Expansive Soils. Texas A&M Univ. Press, College Station, TX, pp. 324–329.

Nobel, C. A. (1966). Swelling measurements and prediction of heave for a lacustrine clay. Can. Geotechn. J. 3(1):32–41.

O'Bannon, C. E., Morris, G. R., and Mancini, F. P. (1976). Electrochemical hardening of expansive clays. Transport. Res. Rec. 593:46–50.

Ofer, Z. (1980). Instruments for laboratory and in situ measurement of lateral swelling pressure of expansive clays. Proc. 4th Int. Conf. Expansive Soils, D. Snethen, ed. Denver, CO, June 1980 1:45–53.

Ofer, Z. (1988). Instruments for laboratory and in-situ determination of lateral soil pressure. Ph.D. dissertation, Univ. of the Witwatersraand, Johannesburg, South Africa.

Olson, R. E., and Langfelder, L. J. (1965). Pore water pressures in unsaturated soils. J. Soil Mech. Found. Div., ASCE, Proc. Paper 9409, July 91 (SM4):127–150.

O'Neill, M. W. (1988). Special topics in foundations. Proc. Geotech. Eng. Div., ASCE National Convention, Nashville, TN, B. M. Das, ed. May, pp. 1–22.

O'Neill, M. W., and Poormoayed, N. (1980). Methodology for foundations on expansive soils. J. Geotech. Eng. Div., ASCE 106 (GT12, December):1345–1367.

Osterberg, J. O. (1979). Failures in exploration programs. Proc. ASCE Specialty Workshop Site Charact. Explor., Evanston, IL, pp. 3–9.

Packard, R. G. (1973). Design of concrete airport pavements. Portland Cement Association, EB050.03P.

Parker, J. C., Amos, D. F., and Kaster, D. L. (1977). An evaluation of several methods of estimating soil volume change. Soil Sci. Soc. Am. J. 41(6):1059–1064.

Pearring, J. R. (1963). A study of basic mineralogical, physical-chemical, and engineering index properties of laterite soils. Dissertation. Texas A&M Univ., College Station, TX.

Peattie, K. R. (1962). Stress and strain factors for three-layer elastic systems. Highway Res. Board Bull. 342.

Petak, W. J., Atkisson, A. A., and Gleye, P. H. (1978). Natural hazards: A building loss mitigation assessment (Final Report). J. H. Wiggens Co. report under NSF Grant ERP-75-09998 (June).

Peter, P., and Martin, R. (1973). The quantitative description of the stress deformation behaviour for expansive soils: 3. A simple psychrometer for routine determinations of total suction in expansive soils. Proc. 3rd Int. Conf. Expansive Soils, Haifa, Israel 2:89–96.

Pickett, G., and Ray, G. K. (1951). Influence charts for rigid pavements. Transact. ASCE. Paper No. 2425, V.116:49–73.

Poor, A. R. (1978). Experimental residential foundation design on expansive clay soils. Final Report, Construction Research Center, Univ. of Texas, Arlington, Rep. No. TR-3-78, June.

Poor, A. R. (1979). Data supplement for remedial measures for houses damaged by expansive clay. Department of Housing and Urban Development, Office of Policy Development and Research. 61 pp.

Porter, A. A. (1977). Mechanics of swelling in expansive clays. M.S. thesis, Colorado State University, Fort Collins, CO.

Porter, A. A., and Nelson, J. D. (1980). Strain controlled testing of soils. Proc. 4th Int. Conf. Expansive Soils, ASCE and Int. Soc. Soil Mech. Found. Eng., Denver, June: 34–44.

Portland Cement Association. (1970). Recommended practice for construction of residential concrete floors on expansive soil, Vol. 2. Portland Cement Association, Los Angeles.

Post, J. L., and Padua, J. A. (1969). Soil stabilization by incipient fusion. Highway Res. Board Special Rep. 103:243–253.

Poulos, H. G., and Davis, E. H. (1980). Pile Foundation Analysis and Design, John Wiley, New York.

PTI. (1980). Design and construction of post-tensioned slabs-on-ground. Post-Tensioning Institute, Phoenix, AZ.

Ralph, T. K., and Nagar, N. (1972). The prediction of heave from laboratory tests. Univ. of Witwatersraand, Johannesburg.

Raman, V. (1967). Identification of expansive soils from the plasticity index and the shrinkage index data. Indian Eng., Calcutta 11 (1):17–22.

Reese, L. C., and Wright, S. J. (1977). Construction of drilled shafts and design for axial loading. Drilled Shaft Design and Construction Guidelines Manual, Vol. 1. U.S. Department of Transportation, Federal Highway Administration, Office of Research and Development, Washington, D.C., July.

Richards, B. G. (1966). The significance of moisture flow and equilibria in unsaturated soils in relation to the design of engineering structures built on shallow foundations in Australia. Symp. Permeability Capillarity, ASTM, Atlantic City, NJ.

Richards, B. G. (1967). Moisture flow and equilibrium in unsaturated soils for shallow foundations. Permeability and Capillarity of Soils, ASTM STP 417, pp. 4–34.

Richards, L. A. (1928). The usefulness of capillary potential to soil moisture and plant investigators. J. Agric. Res. 37:719–742.

Richards, L. A., and Ogata, G. (1958). Thermocouple for vapor pressure measurement in biological and soil systems at high humidity. Science 128:1089–1090.

Richards, S. S. (1965). Soil Suction Measurements with Tensiometers. Methods of Soil Analysis, C. A. Black, D. D. Evans, J. L. White, L. E. Ensminger, and F. E. Clark, eds. Agronomy 9:153–163.

Riggle, F. R. (1978). Soil water potential determination with thermocouple psychrometers. Univ. of Minnesota, St. Paul, Master of Agriculture Integrating Paper.

Robnett, Q. L., Jamison, G. F., and Thompson, M. R. (1972). Stabilization of deep soil layers. Air Force Weapons Laboratory, Kirtland Air Force Base, New Mexico. Tech. Rep. AFWL-TR-71-90.

Russam, K. (1965). The prediction of subgrade moisture conditions for design purposes. Moisture Equilibria and Moisture Changes Beneath Covered Areas. A Symposium in Print. Butterworths, Australia, pp. 233–236.

Sampson, E., Schuster, R. L., and Budge, W. D. (1965). A method of determining swell potential of an expansive clay. Concluding Proc. Engineering Effects of Moisture Changes in Soils. Int. Res. Eng. Conf. Expansive Clay Soils. Supplementing the Symposia in Print. Texas A&M Univ. Press, College Station, TX, pp. 255–275.

Schafer, W. M., and Singer, M. J. (1976). Influence of physical and mineralogical properties of swelling of soils in Yolo County, California. Soil Sci. Soc. Am. J. 40(4):557–562.

Schmertmann, J. H. (1955). The undisturbed consolidation behavior of clay. Transact. ASCE 120:1201–1227.

Schmertmann, J. H. (1974). Guidelines for design using CPT data. Prepared for Fugro-Cesco, Leischendam, The Netherlands.

Schmertmann, J. H. (1975). In situ measurement of shear strength. ASCE Geotechnical

Engineering Division Specialty Conference on in Situ Measurement of Soil Properties, North Carolina State Univ., Raleigh, June 1–4, 1975, State-of-the-Art Report.

Schmertmann, J. H. (1976). Interpreting the dynamics of the standard penetration test. Engineering and Industrial Experiment Station Research Report D-636, Predicting the q/d/c/u/N Ratio. Dept. of Civil Engineering, Univ. of Florida, Gainesville, FL.

Schmertmann, J. H. (1983). Private communication.

Schneider, G. L., and Poor, A. P. (1974). The prediction of soil heave and swell pressures developed by an expansive clay. Research Rep. TR-9-74, Construction Research Center, Univ. of Texas, Arlington.

Sealy, C. O. (1973). The current practice of building lightly loaded structures on expansive soils in the Denver metropolitan area. Proc. Workshop Expansive Clays and Shales in Highway Design and Construction, Univ. of Wyoming, Laramie, May 1:295–313.

Seed, H. B., Mitchell, J. K., and Chan, C. K. (1962a). Studies of swell and swell pressure characteristics of compacted clays. Highway Res. Board Bull. 313:12–39.

Seed, H. B., Woodward, R. J., Jr., and Lundgren, R. (1962b). Prediction of swelling potential for compacted clays. J. Soil Mech. Found. Div., Am. Soc. Civil Eng. 88 (SM3):53–87.

Sherry, G. P. (1982). Constitutive relationships for unsaturated uranium mill tailings. Masters thesis, Colorado State University, Fort Collins, CO.

Shiflett, M. M. (1974). The effect of recycled rubber upon lightly loaded structures founded on expansive clay soils. M.S. thesis, University of Texas at Arlington. 76 pp.

Sibley, J. W., and Williams, D. J. (1990). A new filter material for measuring soil suction. Geotech. Test. J. ASTM 13(4, Dec.):375–380.

Skempton, A. W. (1953). The colloidal activity of clays. Proc. 3rd Int. Conf. Soil Mech. Found. Eng., Switzerland. V.1:57–61.

Skinner, L. W. (1973). Discussion. Proc. Workshop Expansive Clays and Shales in Highway Design and Construction, Vol. 2. Federal Highway Administration, Denver, CO, p. 294.

Slack, D. C. (1975). Modeling the uptake of soil water by plants. Ph.D. dissertation, University of Kentucky, Lexington.

Smith, A. W. (1973). Method for determining the potential vertical rise, PVR. Proc. Workshop Expansive Clays and Shales in Highway Design and Construction, Univ. of Wyoming, Laramie, May 1:189–205.

Smith, R. E. (1973). California's general experiences. Proc. Workshop Expansive Clays and Shales in Highway Design and Construction, Vol. 2. Federal Highway Administration, Denver, CO, pp. 245–249.

Snethen, D. R. (1979a). Technical guidelines for expansive soils in highway subgrades. U.S. Army Eng. Waterway Exp. Sta., Vicksburg, MS, Rep. No. FHWA-RD-79-51.

Snethen, D. R. (1979b). An evaluation of methodology for prediction and minimization of volume change of expansive soils in highway subgrades. Vol. I, Research Rep., U.S. Army Eng. Waterway Exp. Sta., Vicksburg, MS, Rep. No. FHWA-RD-79-49.

Snethen, D. R., and Johnson, L. D. (1980). Evaluation of soil suction from filter paper. Geotech. Lab., U.S. Army Eng. Waterway Exp. Sta., Vicksburg, MS, Misc. Paper No. 6L-80-4.

Snethen, D. R., Townsend, F. C., Johnson, L. D., Patrick, D. M., and Vedros, P. J. (1975). A review of engineering experiences with expansive soils in highway subgrades. Fed. Hwy. Adm., U.S.D.O.T., Rep. No. FHWA-RD-RD-75-48, NTIS No. AD A020309.

Snethen, D. R., Johnson, L. D., and Patrick, D. M. (1977). An evaluation of expedient methodology for identification of potentially expansive soils. Soils and Pavements Lab., U.S. Army Eng. Waterway Exp. Sta., Vicksburg, MS, Rep. No. FHWA-RE-77-94, NTIS PB-289-164.

Spanner, D. C. (1951). The Peltier effect and its use in the measurement of suction pressure. J. Exp. Bot. 2:145–168.

Statement of the Review Panel. (1965). Engineering concepts of moisture equilibria and moisture changes in soils. Moisture Equilibria and Moisture Changes in Soils Beneath Covered Areas. Symposium, Soil Mechanics Section, Commonwealth Scientific and Industrial Research Organization, G. D. Aitchison, ed. Butterworths, Australia, pp. 7–21.

Steinberg, M. L. (1981). Deep vertical fabric moisture barriers in swelling soils. Transport. Res. Board 790:87–94.

Stocker, P. T. (1972). Diffusion and diffuse cementation in lime and cement stabilised clayey soils—physical aspects. Australian Road Research Board, Proc. 6th Conf. 6(pt. 5):235–285.

Sullivan, R. A., and McClelland, B. (1969). Predicting heave of buildings on unsaturated clay. Proc. 2nd Int. Res. Eng. Conf. Expansive Clay Soils. Texas A&M Univ. Press, College Station, TX, pp. 404–420.

Sweeney, D. J. (1982). Some in situ soil suction measurements in Hong Kong's residual soil slopes. Proc. 7th Southeast Asian Soc. Geotech. Eng. Conf. Vol. I, Hong Kong, 91–106.

Taylor, D. W. (1948). Fundamentals of Soil Mechanics. John Wiley, New York.

Teng, T. C. P., and Clisby, M. B. (1975). Experimental work for active clays in Mississippi. Transport. Eng. J. ASCE 101(TEI):77–95.

Teng, T. C. P., Mattox, R. M., and Clisby, M. B. (1972). A study of active clays as related to highway design. Research and Development Division, Mississippi State Highway Dept., Engineering and Industrial Research Station, Mississippi State University, MSHD-RD-72-045, 134 pp.

Teng, T. C. P., Mattox, R. M., and Clisby, M. B. (1973). Mississippi's experimental work on active clays. Proc. Workshop on Expansive Clays and Shales in Highway Design and Construction, Univ. of Wyoming, Laramie, May 2:1–17.

Texas State Department of Highways and Public Transportation. (1970). Manual of Testing Procedures, 100-E Series, Soils Section, Texas State Dept. of Highways and Public Transportation, April.

Thompson, M. R. (1966). Lime reactivity of Illinois soils. ASCE J. Soil Mech. Found. Div. 92 (SM5):67–92.

Thompson, M. R. (1968). Lime stabilization of soils for highway purposes. December 1968. Final Summary Rep. Civil Engineering Studies, Highway Series No. 25, Illinois Cooperative Highway Research Program, Project IHR-76, 26 pp.

Thompson, M. R. (1969a). Lime stabilization: Deep flow style. Road Streets, March.

Thompson, M. R. (1969b). Mixture design for lime-treated soils. Civil Engineering Studies, Highway Engineering Series 26, Illinois Cooperative Highway Research Program. University of Illinois, Urbana. 24 pp.

Thompson, M. R., and Robnett, Q. L. (1976). Pressure-injected lime for treatment of swelling soils. Transport. Res. Rec. 568:24–34.

Thornwaite, C. W. (1948). An approach toward a rational classification of climate. Geograph. Rev. 48(1):55–94.

U.S. Bureau of Reclamation. (1974). Earth Manual, 2nd ed. U.S. Department of the Interior, Water Resources Technical Publication, Denver, CO.

van der Merwe, D. H. (1964). The prediction of heave from the plasticity index and the percentage clay fraction of soils. Civil Eng. South Africa 6:103–107.

van der Merwe, D. H., Hugo, F., and Steyn, A. P. (1980). The pretreatment of clay soils for road construction. Proc. 4th Int. Conf. Expansive Soils, Vol. 1. ASCE, Denver 4:361–382.

Van London, W. J. (1953). Asphalt membranes—their service record on Gulf freeway fills. Roads Streets 96(4):104.

Walsh, P. R. (1978). The analysis of stiffened rafts on expansive clays. Tech. Paper No. 122 (2nd series), C.S.I.R.O. Division of Building Research, Australia.

Westergaard, H. M. (1927). Theory of concrete pavement design. Proc. Highway Res. Board.

Whittig, L. D. (1964). X-ray diffraction techniques for mineral identification and mineralogical composition. Methods of Soil Analysis, American Society of Agronomy Monograph, No. 9, Part I, Ch. 49.

Wiggens, J. H., Slossan, J. E., and Krohn, J. P. (1978). Natural hazards: Earthquake, landslide, expansive soil. J. H. Wiggins Co. report for National Science Foundation under Grants ERP-75-09998 (Oct.) and AEN-74-23993.

Wilkes, P. F. (1970). The installation of piezometers in small diameter boreholes. Geotechnique XX(3).

Williams, A. A. B. (1965). The deformation of roads resulting from moisture changes in expansive soils in South Africa. Moisture Equilibria and Moisture Changes in Soils Beneath Covered Areas, G. D. Aitchison, ed. Symposium, Soil Mechanics Section, Commonwealth Scientific and Industrial Research Organization. Butterworths, Australia, pp. 143–155.

Winterkorn, H. F. (1975). Soil stabilization. Chapter 8. Foundation Engineering Handbook. Van Nostrand Reinhold, New York.

Wise, J. R., and Hudson, W. R. (1971). An examination of expansive clay problems in Texas. Center for Highway Research, Univ. of Texas, Austin, Res. Rep. 118-5, July.

Witczak, M. W. (1972). Design of full-depth asphalt airfield pavements. Proc. 3rd Int. Conf. Structural Design of Asphalt Pavements, London.

Wong, H. Y., and Yong, R. M. (1973). A study of swelling and swelling force during unsaturated flow in expansive soils. Proc. 3rd Int. Conf. Expansive Soils, Haifa, Israel 1:143–151.

Woodburn, J. A. (1974). Design of foundations on expansive clay soils; current design procedures and practice. Rep. No. Z/48. Industrial Research Institute of S. Australia.

Woods, K. B., Berry, D. S., and Goetz, W. (1960). Highway Engineering Handbook. McGraw Hill, New York.

Woodward-Clyde-Sherard and Associates. (1968). Final report of field survey: Remedial methods applied to houses damaged by high volume-change soils. Federal Housing Administration. FHA Contract H-799. 94 pp.

Wray, W. K. (1978). Development of a design procedure for residential and light commercial slabs-on-ground constructed over expansive soils. Ph.D. dissertation, Texas A&M Univ., College Station TX.

Wray, W. K. (1980). Analysis of stiffened slabs-on-ground over expansive soils. Proc. 4th Int. Conf. Expansive Soils, Denver, CO 1:558–581.

Wray, W. K. (1984). The principle of soil suction and its geotechnical engineering applications. Proc. 5th Int. Conf. on Expansive Soils, Adelaide, S. Australia 114–118.

Wright, P. J. (1973). Slurry pressure injection tames expansive clays. ASCE, Civil Engineering, October.

Wroth, C. P., and Hughes, J. M. O. (1983). An instrument for the in situ measurement of the properties of soft clays. Eighth Int. Conf. Soil Mech. Found. Eng., Moscow. V.1:487–494.

Yoder, E. J., and Witczak, M. W. (1975). Principles of Pavement Design, 2nd ed. John Wiley, New York.

Yoshida, R. T., Fredlund, D. G., and Hamilton, J. J. (1983). The prediction of total heave of a slab-on-grade floor on regina clay. Can. Geotech. J. 20, No. 1:69–81.

INDEX

255